God's Signature in the Stars

— Volume One —

God's Signature in the Stars

— Volume One —

Epic Discoveries in Astronomy

Bruce J. Patterson

RESOURCE *Publications* · Eugene, Oregon

GOD'S SIGNATURE IN THE STARS, VOLUME ONE
Epic Discoveries in Astronomy

Resource Publications
An Imprint of Wipf and Stock Publishers
199 W. 8th Ave., Suite 3
Eugene, OR 97401

www.wipfandstock.com

PAPERBACK ISBN: 978-1-6667-6829-9
HARDCOVER ISBN: 978-1-6667-6830-5
EBOOK ISBN: 978-1-6667-6831-2

VERSION NUMBER 07/08/24

To my beautiful bride, Lynne

Contents

Preface

DID GOD INSCRIBE HIS name upon the stars? Can his signature be identified by way of "signs" in the heavens? Indeed, God—the celestial artist-in-residence—"autographs" his starry host. He tells us in Psalm 19 and Psalm 8: "The heavens declare the glory of God; And the firmament shows His handiwork" (Ps 19:1). "When I consider Your heavens, the work of Your fingers, The moon and the stars, which You have ordained . . ." (Ps 8:3). Accordingly, *God's Signature in the Stars* is based on these cosmic confirmations. God is the author of biblical astronomy. In the book of Genesis chapter 2:18–20, we learn God entrusted Adam with naming "all the cattle, the birds of the air, and every beast of the field." Yet, God created and named the stars for his own special purpose—to reveal his eternal message of love to mankind.

God's record of astronomy is one of the oldest sciences and one of man's earliest means of communication. Indeed, astronomy is a universal language. Ancient sailors determined their direction based on the position of "starry hosts." More than a celestial compass, however, as we connect the dots of these starry hosts to the Bible and history, we discover timeless truths.

My study of biblical astronomy began more than thirty years ago after being introduced to the subject by my friend and associate pastor of my church, James R. Bledsoe. Shortly thereafter, I became an avid viewer of Dr. D. James Kennedy's *The Coral Ridge Hour* and an enthusiastic listener to Dr. Kennedy's *Truth in Action* radio broadcast. Dr. Kennedy's book *The Real Meaning of the Zodiac* holds a place of honor in my bookcase. As a mentor in-kind, Dr. Kennedy's gracious and informative responses to my questions in our letters of correspondence are invaluable. Through the years, Mr. Kenneth C. Fleming, faculty emeritus at Emmaus Bible College and author of *God's Voice in the Stars*, has been tremendously supportive of my work and during the process of writing this book.

God's "signature" in the stars is hidden in plain sight. Star names have been preserved in the ancient languages of Hebrew, Aramaic, Arabic, and several other languages. I explore these ancient star names found in the twelve major constellations—including the thirty-six minor (decans/parts) constellations.

Through Scripture, biblical commentary, star maps, definitions, comparison charts, and ancient languages, my work defines and reveals the significance of these star names; confirming "the heavens declare the glory of God," the Jewish Messiah, Yeshua Hamashiach, known as Jesus Christ, as translated in the Greek language.

The Bible verifies God as the author of star names: "He counts the number of the stars; He calls them all by name" (Ps 147:3). To my knowledge—and to date—a modern comprehensive analysis of these star names, based on their ancient languages, has never been so fully documented and made available to the world.

Following biblical principles, I hold fast to Scripture in my analysis of star names and their meanings:

"In the mouth of two or three witnesses shall every word be established" (2 Cor 13:1). And "prove all things; hold fast that which is good" (1 Thess 5:21 KJV). In this regard, I realize we can understand much of the Bible without astronomy, but we cannot fully understand astronomy without the Bible. The Bible is our primary source for interpreting biblical astronomy, as the classic Psalm says: "The heavens declare the glory of God; and the firmament shows His handiwork. Day unto day utters speech, and night unto night reveals knowledge. There is no speech nor language where their voice is not heard. Their line has gone out through all the earth, and their words to the end of the world" (Ps 19:1–4).

God's Signature in the Stars addresses the urgency of the hour through the gospel message in the constellations and biblical prophecy—including never-before-recognized celestial discoveries. Recognizing that God is the author of prophetic "poetry," as displayed in the sky, we realize that astronomy is artfully woven throughout biblical text.

One of the major themes of astronomy and the Bible is found in the first prophecy regarding the promised "Seed of the Woman" (the virgin-born Messiah) who would crush the serpent's head (Gen 3:15). This first gospel defined as "protoevangelium," combines two Greek words: *protos*, meaning "first," and *evangelion*, meaning "good news" or "gospel." Several years ago, as I contemplated the scenes of the cross, I realized that the constellation Hercules, who symbolizes the Messiah, was highlighted by God—during Christ's crucifixion—as crushing the head of Draco, the Dragon (symbolizing Satan) as they descend and disappear beneath the western horizon.

The Bible documents darkness as "covering the land" from 12:00 noon to 3:00 p.m. during the crucifixion of Christ. It appears God "turned down the light," to feature his "screenplay" in the sky. Were the stars and constellations visible during those three hours of darkness as Christ hung on the cross, thus portraying and pointing to the fulfillment of their heavenly message on Earth (real-time)? Regardless, they were present in the "backdrop" of the sky.

At 12:00 noon (the fourth of his seven-hour crucifixion), Leo, the Lion (representing Yeshua, the Lion of the Tribe of Judah), ascended in the east, while Draco, the Dragon (representing Satan), descended in the west. Could these celestial scenes tell the story of the first messianic prophecy, as foretold in Gen 3:15? Indeed, they do—and much more!

The constellation Aries, the Ram (depicting Yeshua, the Passover Lamb), descended beneath the western horizon at precisely 7:00 p.m., after the Messiah was placed in the tomb. Aries reemerges in the eastern sky before sunrise in the early morning hours during this time of the year. On resurrection morning, Sunday, April 29, AD 31, Aries rose at approximately 4:30 a.m. Simultaneously, the constellation Perseus, the Breaker (symbolizing Jesus, as warrior), is rising from "the dark domain" with the sun aglow at his feet, symbolically clutching Satan's head tightly in his left hand.

These cosmic scenes indicate that our Creator calibrated the constellations to portray and confirm, real-time, the death, burial and resurrection of the Messiah and his victory over Satan! The canopy of constellations in the dome of the sky can be likened to actors in a screenplay performing their roles in God's theater in the sky. These "roles" are particularly meaningful on the Lord's biblical feast days.

The feasts of the Lord are called holy convocations. The Hebrew word *convocations* מִקְרָא/*miqrâ'*, means "something called out; a rehearsal" (Lev 23:1, 2).

Fact: The Creator calibrated the cosmos and arranged the constellations in their fixed order.

Explanation: God synchronized the constellations with the biblical feast days. He features specific constellations during his appointed times of the year.

Conclusion: The constellations portray and confirm the death, burial, and resurrection of the Messiah during the annual spring feasts. During the fall feasts the constellations signify a rehearsal of the second coming of the Messiah. Do these signs appear according to God's providence? Do they appear by chance? May my book encourage you to consider, that as the Sun rises and sets at God's appointed time, and the cycles of the Moon are calibrated by the Creator, so goes the course of the constellations as viewed from Earth. Bear in mind, there is no word in the Hebrew language for "coincidence."

The Bible teaches that one can come to faith in God, and his Son, Yeshua, the Messiah—(Jesus Christ) through understanding his celestial message in the stars.

The apostle Paul says: "[17] So then faith *comes* by hearing, and hearing by the word of God. [18] But I say, have they not heard? Yes indeed: 'Their sound has gone out to all the earth, and their words to the ends of the world'" (Rom 10:17, 18). In verse 18, the apostle is quoting Ps 19:4 in reference to the revelation God has given man through biblical astronomy.

The Bible declares the names of several constellations and stars in their original (Hebraic) language. Other constellations are alluded to by writers of Scripture. Although many of the constellations and stars are not directly mentioned in the Bible, an occasional phonetic derivative of their names can be found. I include such biblical references as examples of these derivatives and how they are used.

God painted his "picture gallery" in the sky. His imagery in the constellations conveys the stories and parables of his timeless gospel message.

The following revelations found within the pages of *God's Signature in the Stars* serve as an evangelistic tool:

1. Yeshua's (Jesus') story of love—his birth, death, burial, resurrection, present-day ministry, and soon-coming return—is declared in the heavens.

2. Yeshua *is* "the glory of God" as referenced in Ps 19:1; 2 Cor 4:6; and Heb 1:3.

3. "Great are the works of the LORD; they are studied by all who delight in them (Ps 111:2).

4. Yeshua describes himself as "the Aleph and the Tav" referenced in the Hebrew version of Rev 1:8 and 22:12. The aleph and the tav are the first and last letters of the Hebrew alphabet. The aleph and the tav "(אֶת) /et" is the fourth word in Genesis 1:1. The aleph indicates the source of creation. The early Hebrew tav was shaped like a cross (corresponding to the constellation Crux) and is defined as "a mark; by implication, a signature," thus, *God's Signature in the Stars.*

One of my primary purposes for writing this book is to offer you, the reader, an opportunity to experience a richer revelation and relationship with our Maker and Redeemer. I pray you will draw nearer to the Lord as you come to know him more deeply.

> And this is eternal life, that they may know You, the only true God, and Jesus Christ whom You have sent. (John 17:3)

Through the years, I have written and arranged on classical guitar a series of "star songs" based on the Mazzaroth, the biblical Hebrew word for "constellations." My music, prophetically and poetically, magnifies the message of the constellations and stars in correlation with the Bible.

—Bruce J. Patterson

Acknowledgments

ABOVE ALL, I THANK God for placing in my heart his desire to explore and explain the Biblical meaning of his amazing Mazzaroth.

I wish to express my heartfelt gratitude to my excellent wife, Lynne—the love of my life, my Proverbs 31 wife—for her love and support, musical influence and literary knowledge; to my church family; to my friend and fellow-astronomer, Tim Mitchell; Kenneth C. Fleming, author of *God's Voice in the Stars*; to the saints who have gone before me, having devoted their lives to the truth of the gospel as found in biblical astronomy: Frances Rolleston,[1] E. W. Bullinger,[2] Joseph A. Seiss,[3] and Dr. D. James Kennedy;[4] to Paul and Linda Holland, Pastor Terry Bailey, Pastor Michael and Margaret Duke, Kris and Raili Robbie, Alec and Sharon Doren, Jim and Sandra Pinkoski, David Rives and family, Dr. Chuck and Therese Thurston, Rob Frazier, Byron Spradlin.

1. Rolleston, *Mazzaroth*.
2. Bullinger, *Witness of the Stars*.
3. Seiss, *Gospel in the Stars*.
4. Kennedy, *Real Meaning of the Zodiac*.

Introduction

ARE THERE CREDIBLE REASONS to consider that many of the constellation and star names were given by inspiration of God? *God's Signature in the Stars* explores these ancient star names and their meanings to reveal the answer.

The Bible states that God named the stars: "He who brings out the starry host one by one and calls forth each of them by name" (Isa 40:26). These starry hosts appear in the night sky, as God's "character actors," performing their roles on his celestial stage. God is the creator, writer, and director of this heavenly drama and cosmic dance. By day and by night, his "starry hosts" flicker above the footlights in God's theater in the sky!

Prior to Adam and Eve witnessing a withering leaf die and drift to the ground, the constellation Aries, the Ram (a pictorial symbol of Yeshua, the Lamb of God), hovered over the earth as a testimony of God's grace.

My research is based on the star names and definitions as documented by Frances Rolleston,[1] E. W. Bullinger,[2] and Joseph A. Seiss.[3] My mission is to help restore these constellations and star names to their original narrative—to the glory of God (Ps 19:1). My analysis confirms or corrects the definitions of these star names contained in the aforementioned authors' works. My analysis is based on Ps 89:2: "For I have said, Mercy shall be built up forever; Your faithfulness You shall establish [כּוּן/ *kûwn*] in the very heavens." The Hebrew verb *establish* is כּוּן/*kûwn*, meaning "be firm, arrange, direct; put right, correct."

First-century Jewish historian Flavius Josephus, in his book *The Works of Flavius Josephus*, claims that the record of astronomy dates back to Seth, the son of Adam: "They [Seth and his family] were the inventors

---

1. Rolleston, *Mazzaroth*.
2. Bullinger, *Witness of the Stars*.
3. Seiss, *Gospel in the Stars*.

of that 'peculiar sort of wisdom' which is concerned with the heavenly bodies and their order."[4] Josephus continues, "God afforded these ancients a longer time of life because of their virtue, and the good use they made of it in astronomical and geometrical discoveries."[5] After the flood, evidence reveals that the gospel in astronomy was passed down through Noah's son, Shem (whose name literally means the "name").

Over the last four thousand years, however, many original star names have been lost or distorted through mythology, astrology, and secularism. God's word forbids the unscientific practice of horoscope/astrology, and worship of the creation (Isa 47:12–14; Rom 1:25). God did not design the starry heavens as a means by which to explain or predict human behavior. God designed the starry heavens to reveal his glorious gospel to humanity (Gen 1:14; Ps 19; Rom 10:17,18).

To ensure the credibility of constellations and their star names, it is imperative that research is conducted through the study of their ancient languages. The ancient Semitic languages—Hebrew, Aramaic, Akkadian, Arabic, and Assyrian—define star name meanings more accurately than non-Semitic languages. Therefore, the emphasis of my research stems from the ancient Semitic languages. Occasionally, words have been altered in translation over the centuries. Therefore, I employ etymology, *the study of the origin and evolution of a word-meaning across time.* My objective is to document star names based on their original meaning and overall theme.

The preflood antediluvians may have spoken Hebrew in another vernacular. Most of the ancient names recorded in the first ten chapters of Genesis are of Hebrew origin.

Noah and his family apparently carried the science of biblical astronomy with them on the ark. Over the course of time, they communicated the story of the stars, in the Hebrew language, to the new civilization after the flood. After the tower of Babel dispersion, many people groups carried with them the general teaching of the Mazzaroth/zodiac as they migrated throughout the earth, thus, forming nations. This explains why there must be a common source and that this identifiable common source is the God of the Bible.

Through research, I discovered that many of the Arabic star names may be Hebraic in origin. *God's Signature in the Stars* features lexicon

---

4. Josephus, *Works of Josephus*, 1.2.68–71.

5. Josephus, *Works of Josephus*, 1.3.106.

sources and references to substantiate the definitions of these star names. Hebrew is the predominate root language.

Ancient languages, as referenced in my research, include Hebrew, Aramaic, Arabic, Akkadian, Assyrian, Phoenician, Persian, Sumerian, Sanskrit, Hindi, Tamil, Turkish, Greek, Latin, Egyptian Arabic, Egyptian, Ethiopian, and Chinese.

The book of Job is considered to be the oldest book in the Bible and in the world. God uses the word מַזָּרָה/*mazzârâh* in Job 38:32 when referring to the constellations. The word *mazzârâh* is from נָזַר/*nâzar*, meaning "to dedicate, consecrate, separate." The word נָזַר/*nâzar* is used in reference to the Nazarite who is separated unto God, as were Samuel and Samson. Hence, the Mazzaroth is consecrated and set apart by God. Indeed, the Mazzaroth proclaims the old, old story of Jesus and his love.

The Mazzaroth (signs in the heavens) are known in Greek etymology as the zodiac. There are twelve major constellations of the Mazzaroth/zodiac—including their thirty-six associated constellations.

The word *zodiac* comes from the Latin word *zōdiacus*, whose etymology originates in the Greek word *kýklos*, "circle," and ζωο/*zōo*, meaning "animal," from which our English word "zoo" is derived. Zodiac, therefore, carries the meaning "circle of animals." Notably, many constellations are named after animals.

The *ecliptic* is the apparent path the Sun travels in the course of a year. The annual *ecliptic* circuit is plotted through the "great circle in the sky" encompassing the twelve major constellations. The Earth requires one year to orbit the Sun. The Sun takes the same length of time to make a complete circuit of the ecliptic. This process repeats annually in slightly more than 365 days. Viewed from Earth, the Sun appears to traverse the canopy of constellations. The planets of our solar system, moreover, travel in orbits closely along the ecliptic.

Johannes Kepler, the seventeenth-century German Christian astronomer, used mathematics to calculate the path of the planets. Kepler's findings reveal that planets do not travel in circles but in *ellipses*. Kepler's First Law says: "Each planet's orbit about the Sun is an ellipse. The Sun's center is always located at one focus of the orbital ellipse. The planet follows the ellipse in its orbit, meaning that the planet to Sun distance is constantly changing as the planet goes around its orbit."[6]

---

6. NASA, "Orbits and Kepler's Laws."

Kepler's math is utilized in today's astronomy software. *Stellarium* software, for example, enables people to use their home computer as a virtual planetarium. Stellarium presents a realistic "sky" in 3D format comparable to the naked eye, binoculars, or telescopes. Stellarium software calculates the positions of the Sun, Moon, planets, and stars while displaying the sky as it appears based on time zone and location.

The Creator did not "fling" the stars and constellations into our Milky Way Galaxy. He positioned them deliberately—with design and purpose—as a tool of celestial education.

The book of Job (26:13) declares: "By His Spirit He adorned the heavens; His hand pierced the fleeing serpent." The Hebrew word for adorned is שִׁפְרָה/*shiphrâh*, meaning "brightness, or garnish," from the root שָׁפַר/*shâphar*, "to glisten." This concept of "brightness" applies to the brilliancy of design, order, and sound, similar to the word שׁוֹפָר/*shôwphâr*, "shofar, trumpet."

"His hand pierced the fleeing serpent" corresponds to the constellation Hydra (symbolic of Satan) fleeing from Leo, the Lion, symbolic of Yeshua, the Lion of the tribe of Judah. The constellations present a tangible picture of Messiah's victory over the enemy and our participation with him in his kingdom reign.

C. S. Lewis said: "I take Psalm 19 to be the greatest poem in the Psalms and one of the greatest lyrics in the world."[7]

"The heavens declare [סָפַר/*sāpar*] the glory of God" (Ps 19:1). The Hebrew word for *declare* is סָפַר/*sāpar*, meaning "to score or record, to recount, celebrate, count, scribe, shew forth, speak, talk, tell (out), writer." The semitic root *spr* means "to measure, count, scribe."

God uses the word *declare* (סָפַר/*sāpar*) in his promise to Abraham: "Then He brought him [Abraham] outside and said, 'Look now toward heaven, and count [סָפַ/*sāpar*] the stars if you are able to number [סָפַר/*sāpar*] them.' And He said to him, 'So shall your descendants be'" (Gen 15:5).

God asked Abraham if he could *count, declare, tell (the story in) the stars.* The heavens declare the glory of God—and the glory of God is Yeshua! Each constellation plays a major role in His-story.

When we compare God's story in the stars to the Bible, we observe a unique synergy. Both revelations are inspired by God working through human beings. The heavens *and* the Bible reveal God as both Creator and

---

7. Lewis, *Reflections*, 63.

Redeemer. "Yes, the works of the Lord are great, studied by all who have pleasure in them" (Ps 111:2).

The Father of Lights is the one who made the greater light (Sun) to rule by day and the lesser light (Moon) to rule by night. He also made the stars (Gen 1). Above all, the primary purpose for the starry universe within our galaxy is for "signs and seasons."

"Then God said, "Let there be lights in the firmament of the heavens to divide the day from the night; and let them be for signs [אוֹת/*oth*] and seasons [מוֹעֵד/*moed*], and for days and years" (Gen 1:14). The heavens determine our days and years, *and* they remain for signs and seasons. The Hebrew word for *signs* is אוֹת/*oth*, defined as "a signal (literally or figuratively), a flag, beacon, monument, omen, evidence, etc.—mark, miracle, (en-) sign, token." The Hebrew word for *seasons* is מוֹעֵד/*moed*, which means "an appointment"; a fixed time or "season"; specifically, a festival; conventionally, one year. "Season," however, does not refer to the annual "four seasons," but rather to the Lord's feast days. The Lord's biblical feast days include the weekly Shabbat (Sabbath) and the annual feast days: Passover, Unleavened Bread, Firstfruits, Pentecost, the Feast of Trumpets, Day of Atonement, and the Feast of Tabernacles. The lunar calendar designates the times and seasons in which to honor the Lord's appointed feast days. God's cosmic clock is precise to the second.

According to the Targum (Jewish Aramaic translations of books of the Hebrew Bible) the children of Issachar were all astronomers:

> And the sons of Issachar, who had understanding to know the times, and were skilled in fixing the beginnings of years, the commencement of months, and the intercalation of months and years; skillful in the changes of the moon, and in fixing the lunar solemnities to their proper times; skillful also in the doctrine of the solar periods; astronomers in signs and stars, that they might show Israel what to do; and their teachers were two hundred chiefs of the Sanhedrin: and all their brethren excelled in the words of the law, and were endued with wisdom, and were obedient to their command.

It appears that in their wisdom, experience, and skill, their brethren had the fullest confidence; and nothing was done but by their direction and advice (based on 1 Chr 12:32).[8]

---

8. Tg. 1 Chronicles 12:32.

Yeshua calls us to be alert and to watch for the "signs in the heavens" as we approach his return:

> And there will be signs in the sun, in the moon, and in the stars; and on the earth distress of nations, with perplexity, the sea and the waves roaring; men's hearts failing them from fear and the expectation of those things which are coming on the earth, for the powers of the heavens will be shaken. Then they will see the Son of Man coming in a cloud with power and great glory. Now when these things begin to happen, look up and lift up your heads, because your redemption draws near. (Luke 21:25–28)

The ultimate sign of his coming is foretold in the book of Matthew: "Then the sign of the Son of Man will appear in heaven" (Matt 24:30).

God has given each star a unique color spectrum. "For one star differs from another star in glory" (1 Cor 15:41). "His wonders are without number" (Job 9:10).

The redemptive message of the twelve major constellations *begins* with Virgo, the Virgin and *ends* with Leo, the Lion. This "beginning and ending" is expressed through the structural design of the ancient Egyptian sphinx (see "Virgo," chapter 1).

The calendar sequence of the constellations begins with Aries, the Ram and concludes with Pisces, the Fishes. Aries corresponds to Nisan, the first month on the Hebrew calendar. Passover, the fourteenth day of Nisan, corresponds to Yeshua, the Lamb of God. Pisces corresponds to Adar, the twelfth month on the Hebrew calendar and is a prophetic picture of the church age and *the harvest of souls.*

God's revelations to mankind can be likened to a two-edged sword. The first revelation is revealed through the record of *astronomy.* The second revelation is revealed *through* the Bible.

The first revelation expresses, "The *heavens* declare the glory of God" (Ps 19:1). When referring to the heavens, the apostle Paul says, "Faith comes by hearing and hearing by the word of God" (Rom 10:17).

The second revelation, Ps 19:7, "the Torah of the Lord is perfect," represents the TaNaKh, an acronym for the three divisions of the Hebrew Bible:

1. Torah ("Teaching," also known as the Five Books of Moses)

2. Nevi'im ("Prophets'")

3. Ketuvim ("Writings")

This scriptural (second) revelation includes the New Testament writings of the apostles. When referring to the Scriptures, the apostle Paul says, "Faith comes by hearing and hearing by the word of God" (Rom 10:17).

Understanding God's message in the heavens gives us a greater appreciation for our Creator, Redeemer, and his gift of salvation. My prayer for you, beloved, is expressed by the prophet Daniel: "Those who are wise shall shine like the brightness of the firmament, and those who turn many to righteousness like the stars forever and ever" (Dan 12:3).

The Constellation: Virgo

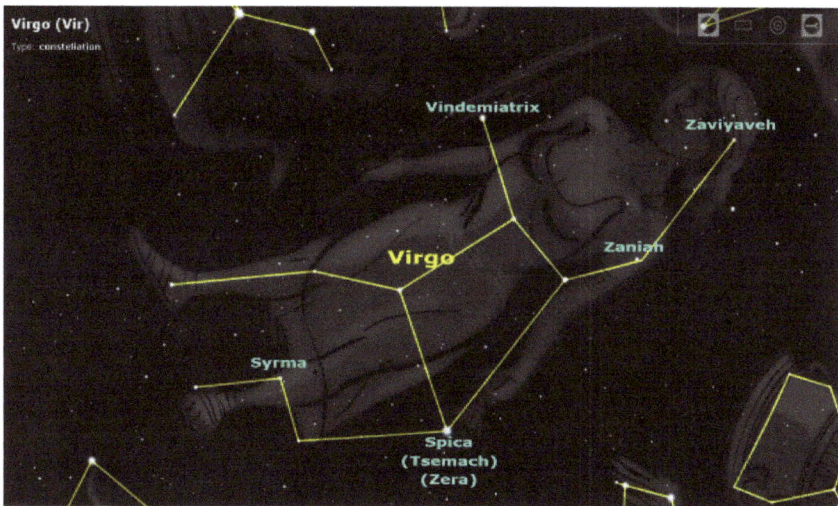

Stellarium image.[1]

THE REDEMPTIVE MESSAGE OF the twelve major constellations begins with Virgo, the Virgin and ends with Leo, the Lion (of Judah). This "beginning and ending" is communicated through the figure of the Egyptian sphinx.

---

1. This and all similar Stellarium images courtesy of stellarium.org and are freely available for republication and reuse.

The Egyptian sphinx is composed of the head of a woman and the body of a lion. The head of the woman represents Virgo, the Virgin—the "beginning." The body of the lion represents Leo, the Lion—the "ending" when Jesus, Yeshua the Lion (of Judah)—returns for his Bride, the church, composed of Jew and Gentile, one in Messiah.

Virgo is the Latin word for "virgin." Similarly, the Latin word *virga* means "branch" or "rod." One of the Messiah's biblical titles is "the Branch." "Behold, the Man whose name is the BRANCH!" (Zech 6:12).

In approximately 750 BC, the Lord inspired Isaiah, the prophet, to direct us to the true meaning of Virgo: "Therefore, the Lord Himself will give you a sign: Behold, the virgin shall conceive and bear a Son, and shall call His name Immanuel" (Isa 7:14).

In Isa 7:14, the word translated as "virgin" is עַלְמָה/*almâh*, meaning "a young woman; a virgin." We know Miriam (Mary) was both עַלְמָה/*almâh*, a young woman and בְּתוּלָה/*bethulah*, a pure virgin. *Bethulah* is the Hebrew name for the constellation Virgo.

"Bethulah" is a prophetic picture of the birth of Messiah—the promised "Seed of the Woman," the Savior of the world. The Lord communicated this life-saving message in his first messianic prophecy as recorded in Genesis: "And I will put enmity between you and the woman, and between your seed and her Seed; he shall bruise your head, and you shall bruise His heel" (Gen 3:15).

Virgo is referenced in the twelfth chapter of the book of Revelation. The woman (Virgo) is described as being clothed with the Sun, with the Moon beneath her feet, and having a crown of twelve stars: "Now a great sign appeared in heaven: a woman clothed with the sun, with the moon under her feet, and on her head a garland of twelve stars. Then being with child, she cried out in labor and in pain to give birth" (Rev 12:1, 2).

In his 1996 book *The Star That Astonished the World*, Dr. Earnest Martin identified Rev 12:1, 2 as an astronomical indicator of the precise time (September 11, 3 BCE) of the Messiah's birth.[2] It was made popular by Rick Larson in his 2007 documentary *The Star of Bethlehem*.

This Rev 12 sign occurred September 11, 3 BC, during the Feast of Trumpets—*when the Sun was in Virgo, the new Moon was beneath her feet, and Jupiter, Venus, and Mercury (including the nine stars of Leo, the Lion), constituted the twelve stars of her crown.*[3] This great celestial sign,

---

2. Martin, *Star That Astonished the World*.

3. Larson, *The Star of Bethlehem*

as recorded in Revelation chapter 12, points to the birth of the Messiah. A three-planet alignment at the head of Virgo, with the new Moon at her feet on Feast of Trumpets, is rare. A similar sign occurred September 23, 2017, on the Feasts of Trumpets and another will occur on Trumpets September 29, 2030.

The Feast of Trumpets is an annual rehearsal for the second coming of the Messiah. Could the Rev 12 sign—which signified the birth of the Messiah—be similar to the sign Yeshua declares will appear before his return? I am proposing that a similar celestial configuration represents *the sign of the Son of Man in Heaven.*

"Then the sign of the Son of Man will appear in heaven, and then all the tribes of the earth will mourn, and they will see the Son of Man coming on the clouds of heaven with power and great glory" (Matt 24:30).

Virgo represents Israel—the country and the Jewish people—giving birth to the Messiah. By rightly dividing prophetic scripture, we can confirm that Miriam, the mother of Yeshua, was an Israelite. Miriam is of the house of David and the tribe of Judah. Yosef (Joseph), although not the biological father of Yeshua, is of the house of David and the tribe of Judah. Miriam would carry Yeshua, the sinless Son of God, in her womb and give birth to the Messiah who, in turn, would bear her sins on the cross.

God chose to reveal himself as a Jewish man in order to redeem mankind. Therefore, the gospel is to be preached to the Jew first and also to the Gentile. "For I am not ashamed of the gospel of Christ: for it is the power of God unto salvation to everyone that believes; to the Jew first, and also to the Greek" (Rom 1:16).

According to the Jewish historian Josephus, Abraham delivered the science of astronomy to the Egyptians.[4] Therefore, the pharaohs and Egyptians understood that the head of the Egyptian sphinx represented Virgo, and the body of a lion represents Leo. But did they perceive that these constellations were designed by the Creator of the universe? Did they know these starry hosts represent the Messiah who would come to Earth in the form of a man from the Hebrew nation?

Did the Hebrews, while laboring and toiling as enslaved people near the sphinx, realize their Deliverer (as represented in Virgo and Leo) would come from the tribe of Judah some fifteen hundred years later? Did Miriam and Yosef, with their little boy, see the Egyptian sphinx as

---

4. Josephus, *Works of Josephus*, "Antiquities of the Jews," 1.166-68.

they fled from King Herod of Judea? Wasn't it King Herod who ordered the genocide of all male children who were two years of age and younger in Bethlehem and in all that region? Could Miriam have known that *she* was the ultimate fulfillment of the constellation Virgo, that *her* little boy was the ultimate fulfillment of the constellation Leo?

Luc Olivier Merson, *Rest on the Flight into Egypt* (1879)
Photograph © Museum of Fine Arts, Boston

Great Sphinx of Giza: The ancient Egyptians understood that the twelve major constellations begin with Virgo and end with Leo. The Sphinx is the oldest known monumental sculpture in Egypt and is commonly believed to have been built by ancient Egyptians of the Old Kingdom during the reign of Pharaoh Khafre (2000–2532 BC). The Sphinx is 240 feet (73 meters) long and 66 feet (20 meters) high. ("Great Sphinx of Giza," Creative Commons, https://commons.wikimedia.org/wiki/File:Great_Sphinx_of_Giza_May_2015.JPG.)

Beneath Virgo and Leo, the constellation Hydra, according to Greek legend, was a gigantic water snake monster having seven to nine heads. The constellation Hydra is alluded to in the twelfth chapter of the book of Revelation: "And another sign appeared in heaven: behold, a great, fiery red dragon having seven heads and ten horns, and seven diadems on his heads. His tail drew a third of the stars of heaven and threw them to the earth. And the dragon stood before the woman who was ready to give birth, to devour her Child as soon as it was born" (Rev 12:3, 4).

In the book of Job, Hydra is referred to as *the fleeing serpent*— "fleeing" from the constellation Leo, the Lion (Jesus the Messiah). "By His Spirit He adorned the heavens; His hand pierced the fleeing serpent [נָחָשׁ בָּרִחַ/*nachash barach*]" (Job 26:13).

Hydra corresponds to (לִוְיָתָן נָחָשׁ בָּרִחַ *Leviathan, the fleeing serpent*, as mentioned by Isaiah, Isa 17:1). Hydra, in keeping with the seven head-ed celestial dragon mentioned in the book of Revelation, is the longest constellation in the sky—encompassing a third of the visible heavens. The seven heads of this celestial dragon symbolize the world empires that have persecuted God's people throughout history.

Draco, the Dragon (representing Satan) is also alluded to in the twelfth chapter of the book of Revelation (Rev 12:7). In the book of Isaiah, Draco is called, "Leviathan, that twisted serpent," לִוְיָתָן נָחָשׁ עֲקַלָּתוֹן/*leviathan aqallathon nachash* (Isa 27:1).

The imagery in the book of Revelation depicts the "dragon" waiting for Virgo, the Virgin, to give birth to the Messiah that he might devour the child. Within the configuration of constellations, Draco, the Dragon, is positioned in the northern region of the sky. This exalted position was symbolically claimed by Satan when Adam and Eve sinned and lost their God-given dominion. Hence, Satan became the god of this world by means of transferred authority (2 Cor 4:4).

Draco's head is pointed toward the Virgin, as if anticipating the birth of the Son of God. An earthly manifestation of this sign occurred when King Herod attempted to "devour the virgin's child [Yeshua, son of

Miriam]" by issuing a decree calling for all male babies, two years of age and younger, to be killed (Matt 2:16–18).

According to the Bible, the Son of God will be born of a virgin. The virgin (Miriam) would be impregnated by the Holy Spirit. This Seed will be enclosed in the holy blood of God within her placenta until his birth. Only the pure blood of God can redeem the people of God.

In Revelation chapter 12, we see the virgin's child caught up to God and his throne. Yeshua has been reigning nearly two thousand years. Yeshua, the Seed of the Woman, will return one day as the Lion of the tribe of Judah. In the meantime, the woman, representing the redeemed (the seed of her offspring), has God's help on Earth as we deal with opposition and hostility waged against us.

When we look into the starry hosts, may we fully appreciate Virgo, the constellation, as God's signature in the stars.

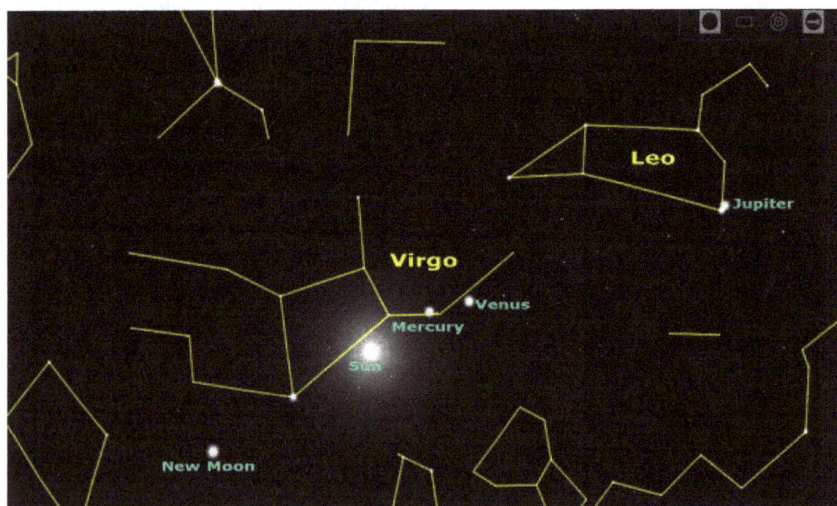

Stellarium image: This astronomical configuration occurred September 11, 3 BC, during the Feast of Trumpets, when the Sun was in Virgo, the new Moon was beneath her feet, and Jupiter, Venus, and Mercury (including the nine stars of Leo, the Lion), constituted the twelve stars of her crown. This great celestial sign, as recorded in Rev 12, points to the birth of the Messiah.

My star song for Virgo is titled "Gloriously Beautiful."

Gloriously Beautiful

(Isa 7:14; Zech 6:12; John 12:24)
By Bruce J. Patterson

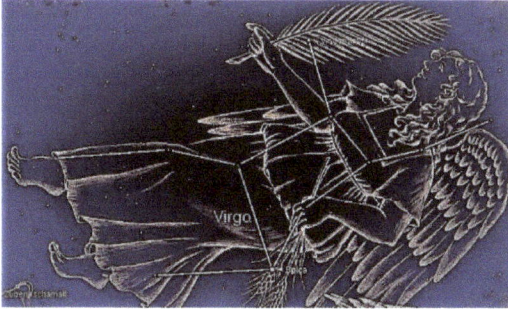

Image courtesy of Starry Night Astronomy, 2023, all rights reserved.

Verse 1
Behold, a virgin shall be with child
And shall bring forth a Son
And they shall call his name
Immanu EL

Chorus 1
Immanu EL, Immanu EL
They shall call his name, Immanu EL
God with us, God with us
We call him Jesus

Verse 2
He, the virgin's seed, stem of Jesse
Ear of wheat
Speak of Messiah's birth, the Son who comes
Branch of Jehovah

Chorus 2
Gloriously beautiful, gloriously beautiful
Jehovah's Branch, Jehovah's Branch
Gloriously beautiful

Bridge
He was born to crush the serpent's head
The grain of wheat to fall and die
Into the earth and bring forth fruit
Chorus 2 and 1

Constellation Name	Translation
Virgo (Latin)	virgin
Bethulah (Hebrew)	virgin
Parthenos (Greek)	the virgin
al-'Athra (Arabic)	the maiden
shì nǔzuò (Chinese)	the virgin constellation
ABSIN (Sumerian)	the seed-furrow
Absinnu (Akkadian)	the seed-furrow
Coume (Coptic)	the Virgo constellation

Enclosed are derivatives and similar words in ancient languages which help clarify the core meaning of this constellation/star name.

Ancient Hebrew/ Aramaic	Phonetic Spelling	Summarized Meaning
Hebrew בְּתוּלָה	bethulah	a pure virgin
Hebrew עַלְמָה	almâh	virgin, young woman
Aramaic ܒ݁ܬ݂ܘܠܬ݁ܐ	B'T,uWLeA	virgin

Ancient Gentile Language	Phonetic Spelling	Summarized Meaning
Latin	virgō	virgin
Greek	parthenos	virgin
Arabic	aḏrā	virgin
Assyrian	btu:l ta	virgin
Persian	batūl	virgin
Akkadian	batultu	virgin

Scripture References

Gen 24:16; Lev 21:14; Deut 22:19; Isa 7:14; Zech 6:12; Matt 1:23; Rev 12:1–2; Matt 24:30

The Stars

Spica

SPICA IS DEFINED IN Latin as "ear of wheat." Held in the virgin's left hand, an insightful explanation of this symbol is stated by Yeshua himself in response to the Greeks coming to Jerusalem to worship during the Feast of Passover. The Greeks asked Philip if they could see Yeshua. Yeshua knew he was born to die according to the scriptures. Through his death, burial, and resurrection, his life would be multiplied through all who receive him as Savior and Lord—the Jewish people first, and then the Gentiles (Rom 1:16). "Most assuredly, I say to you, unless a grain of wheat falls into the ground and dies, it remains alone; but if it dies, it produces much grain" (John 12:24).

Yeshua *is* the grain of wheat; the illustrious Seed of the Woman, the Savior of the world. The children of the kingdom of God are also represented as wheat (Matt 13:24–29).

The Hebrew name for Spica is צֶמַח/*Tsemach*, meaning "branch." The Messiah is often referred to as *Tsemach* or branch. The following scripture is an example of the word "branch" in the Hebrew Bible: "In that day the Branch [*Tsemach*] of the LORD shall be beautiful and glorious. And the fruit of the earth shall be excellent and appealing for those of Israel who have escaped" (Isa 4:2).

Yeshua is the Branch (*Tsemach*) of the LORD! God's people are the glorious offspring of Yeshua. This corresponds to another Hebrew name for this same star, called זֶרַע/*Zera*, meaning "seed." The following is an example of this word in the Hebrew Bible: "Then He brought him [Abram] outside and said, 'Look now toward heaven, and count the stars if you are able to number them.' And He said to him, 'So shall your descendants [*zera*] be'" (Gen 15:5).

When we receive Yeshua as our personal Savior, we become the heavenly seed of Abraham (read Gal 3:15–29; 1 John 3:9). The Lord told Abraham his seed would be like the sand of the seashore. The (sand) represents Abraham's earthly descendants. The Lord also told Abraham his seed would be like the stars of heaven. This (seed) represents those who are born from above (please read Gen 22:17, 15:4; Phil 2:15). When we receive God's gift of salvation, we become a part of the gospel story found in the telling of the stars.

Conventional Star Name/Meaning	Name/Meaning Confirmed or Corrected
Spica—ear of wheat (Latin)	confirmed
Tsemach—branch (Hebrew)	confirmed
Zera—seed (Hebrew)	confirmed
Nefer—beautiful (child) (Egyptian)	confirmed

Enclosed are derivatives and similar words in ancient languages which help clarify the core meaning of this constellation/star name.

Ancient Hebrew/ Aramaic	Phonetic Spelling	Summarized Meanings
Hebrew צֶמַח	*tsemach*	branch
Hebrew זֶרַע	*zera*	seed
Aramaic ܣܘܟܐ	*SaWK'aA*	branch
Aramaic ܙܪܥܐ	*ZaR'aA*	seed

Ancient Gentile Language	Phonetic Spelling	Summarized Meanings
Latin	*spīca*	ear of wheat
Greek	*stáchys*	ear of wheat or other grain
Sanskrit	*zAkhA*	branch
Chinese	*zhī*	branch
Assyrian	*zera*	seed

Ancient Gentile Language	Phonetic Spelling	Summarized Meanings
Hindi	*zīra*	cumin seed
Turkish	*zerre*	grain
Sumerian	*zeru*	seed
Phoenician	*zr*	offspring
Egyptian	*Nefer*	beautiful (child)

Scripture References

John 12:24; Jer 33:15; Isa 4:2; Zech 3:8, 6:12; Gen 15:5; Gal 3:16

Zavijava

The star name Zavijava (pronounced zah-vee-JAH-vah) is a combination of two Hebrew words: *zavi*, צְבִי/*tsev·ē*, meaning "beautiful, glorious," and *java*, from יְהוָה/*Y^ehôvâh*. *Java* or *Jah* is an apparent derivative of *Y^ehôvâh*. According to Dr. Nehemia Gordon, "The Lord's Hebrew name *Y^ehôvâh* is the combination of three words: *yihyeh* 'will be,' *hoveh* 'is' and *hayah* 'was.'"[1]

Yeshua is the ultimate Seed of the Woman (Virgo), the personification of *Y^ehôvâh*. Zavijava, therefore, means "beautiful LORD" in Hebrew. A similar pronunciation and meaning is found in the selective Gentile languages (see chart).

A classic example of the use of these words is found in Isaiah: "In that day shall the branch of the LORD [*Y^ehôvâh*] be beautiful [*tsev·ē*] and glorious, and the fruit of the earth shall be excellent and appealing for those of Israel who have escaped" (Isa 4:2). This ancient prophecy proclaims the fruit of Messiah in unison with the heavens.

The fruit of Messiah is the fruit of the Spirit reproduced in and through the Lord's people here on Earth. It is a beautiful and glorious reality when we produce the fruit of the Spirit. Yeshua said, "By this My Father is glorified, that you bear much fruit; so you will be My disciples" (John 15:8).

---

1. Gordon, "Hebrew Voices #47."

Conventional Star Name/Meaning	Name/Meaning Confirmed or Corrected
Zavijava—gloriously beautiful	confirmed: beautiful LORD

Ancient Hebrew/ Aramaic	Phonetic Spelling	Summarized Meaning
Hebrew צְבִי	*tsev·ē*	beautiful, glorious
Hebrew יָה	*Yâhh, yaw; (Jah)* an abbreviated form of *Y^ehôvâh*	LORD (will be, is, was)
Hebrew יְהֹוָה	*Y^ehôvâh*	LORD (will be, is, was)
Aramaic ܬܫܒܘܚܬܐ	*Tshbwhta*	glory

Enclosed are derivatives and similar words in ancient languages which help clarify the core meaning of this constellation/star name.

Ancient Gentile Language	Phonetic Spelling	Summarized Meaning
Assyrian	*Yi ' hu: wa*	Jehovah
Persian	*Yāhū*	Jehovah
Arabic	*Yahwah*	Jehovah
Turkish	*Yehova*	Jehovah
Tigrinya/Ethiopian	*Yehowa* and *Jehova*	Jehovah
Chinese	*Yēhé huá*	Jehovah
Sanskrit	*Yava*	a stem, any grain of seed or seed corn, barley
Arabic	*ṣabīḥ*	beautiful
Arabic	*ẓaby*	gazelle
Persian	*zībā'ī*	beauty, gracefulness
Assyrian	*zi: wa*	beauty
Sanskrit	*zrI*	beauty
Chinese	*zī*	beauty
Phoenician	*zr*	offspring
Hindi	*rava*	grain
Chinese	*jià*	sow grain

Scripture References

Gen 2:4; Jer 23:5, 33:15; Isa 4:2, 24:16; Zech 3:8, 6:12; Rev 1:8

Vindemiatrix

The star name Vindemiatrix (pronounced vin-de-mee-AY-tricks) is Latin for "vine-harvester" or "grape-harvester." Vindemiatrix is a combination of three Latin words, plus the suffix *trix*. The three words are *vīnī* (wine, figuratively, grapes, grapevine), *dē* (from or out of), and *emia* (suffix meaning blood). Together, we read "wine (or grapes/grapevine) out of blood."

Located in the middle of Vindemiatrix is the phonetic Hebrew word for blood, דָּם/*dâm* (blood, wine) the juice of the grape; figuratively (especially in the plural) bloodshed.

The Messiah—whose blood was not tainted by sin—is the personification of the Seed of the Woman. When we receive God's gift of salvation, we learn that his blood alone provides remission for our sins.

Grapes in the Bible represent fruit produced through a grapevine. The Lord does not say that he is the grape. He says he is the True Vine. In ancient Israel, grapes were the first major crop to ripen. The grape harvest was usually completed before Sukkot, which is the Feast of Tabernacles. Ancient Israel is likened unto a vineyard with a wall and winepress (Isa 5; Matt 20 and 21). The wall represents the wall surrounding Jerusalem. The winepress represents the temple and sacrificial system of the old covenant.

An example of the new covenant biblical concept of the vine and the vinedresser is found in the book of John: "I am the true vine, and My Father is the vinedresser" (John 15).

According to Yeshua's grapevine metaphor, the most important part of one's life is based on the unseen. All humans are likened unto branches connected to a vine. Are your "branches" growing from the rich root system in Messiah, the True Vine, or are they rooted in a false vine? Is your life source in Christ? The Father desires the lifegiving sap of the True Vine to flow into all branches of his vine. He desires that we produce "grapes" of love, humility, and patience called the "fruit of the Spirit." The maturity of this fruit is produced in those who abide in the vine by keeping his word.

Revelation chapter 14 depicts the gathering of grapes. These are "grapes of wrath" and refer to the fruit of the antichrist spirit, with its bitter, sinful nature of hate, pride, and malice.

"Then another angel came out of the temple which is in heaven, he also having a sharp sickle . . . Thrust in your sharp sickle and gather the clusters of the vine of the earth, for her grapes are fully ripe" (Rev 14:17–18).

One of the earliest prophecies regarding the second coming of Messiah is found in Jacob's prophecy to Judah. This prophecy refers to Yeshua, the Messiah, from the lineage of the tribe of Judah: "He washed his garments in wine, and his clothes in the blood of grapes" (Gen 49:11). Many assume this prophecy refers to the wine Judah would enjoy from vineyard grapes. Others presume the passage is consistent with Isa 63:1–6, which describes the day of vengeance associated with the second coming of Messiah. In this passage, "His blood-stained garments" refers to his judgment upon the nations who would attempt to destroy Jerusalem, Israel, and reject his offer of salvation. This imagery, expanded in the book of Revelation, describes the Messiah's robe dipped in blood due to trampling the grapes of wrath.

"He Himself treads the winepress of the fierceness and wrath of Almighty God" (Rev 19:15; read also: Isa 63:1–6, 59:16–17; Rev 14:18–20, 19:11).

The redemptive message of the twelve major constellations begins with Virgo, the Virgin—the birth of Messiah—and ends with Leo, the Lion—the triumphant return. It is interesting to note that the sickle-shaped head of Leo may represent the final harvest. In the parable of the wheat and the tares, Yeshua explains that the harvest takes place at the end of the age (Matt 13:39).

Conventional Star Name/Meaning	Name/Meaning Confirmed or Corrected
Vīndēmiātor—the son, or branch, who comes	corrected: a grape-gatherer; the harbinger of vintage

Enclosed are derivatives and similar words in ancient languages which help clarify the core meaning of this constellation/star name.

Ancient Hebrew/ Aramaic	Phonetic Spelling	Summarized Meaning
Hebrew דָּם	dâm	(blood, wine) the juice of the grape; figuratively (especially in the plural) bloodshed
Aramaic ܕܡܐ	D'iMaA	blood, resemble, liken to, compare

Ancient Gentile Language	Phonetic Spelling	Summarized Meaning
Latin	vīndēmiātor, vindemia, vindemiae	a grape-gatherer
Latin	vindēmia, vindemiae	a grape—gathering, vintage
Latin	vīnus, vīnī	wine, grapes
Latin	vīnea, vīneae	a plantation of vines, vine garden, vineyard, a vine
Latin	vēna, venae	a blood vessel, vein
Arabic	dam	blood
Arabic	dāmin	bloody
Akkadian	damu	blood
Assyrian	dim ma	blood
Phoenician	dm	blood
Persian	dam	blood
Turkish	dem	blood
Amharic/Ethiopian	dem	blood
Amharic/Ethiopian	demi	bloody
Sumerian	dim	vine

Scripture References

Gen 49:11; Isa 5; Rev 14:17–20; Matt 13:39; Matt 20 and 21; John 15

Subilon

The star name Subilon phonetically resembles the Hebrew word *shibbol* or *shibboleth*, meaning "ear of grain; branch."

Subilon may be another name for Spica (in the virgin's left hand) or, it may refer to a star in the virgin's right hand. A derivative example of the word *subilon* is found in the book of Genesis: "And he [Joseph] slept and dreamed the second time: and, behold, seven ears of corn [*Shibbol*] came up upon one stalk, rank and good" (Gen 41:5).

A profound example of the meaning of "ear of wheat" is stated by Yeshua: "Most assuredly, I say to you, unless a grain of wheat falls into the ground and dies, it remains alone; but if it dies, it produces much grain" (John 12:24).

Yeshua defines himself as the "grain of wheat" (*shibbol*), as pictured in Virgo. He is the illustrious Seed of the Woman; the Savior of the world. Wheat represents the children of the kingdom of God (Matt 13:24–29).

Interestingly, the word *subilon* was used as a phonetic test, as recorded in the book of Judges: "And when any Ephraimite who escaped said, 'Let me cross over,' the men of Gilead would say to him, 'Are you an Ephraimite?' If he said, 'No,' then they would say to him, 'Then say, "Shibboleth"!' And he would say, 'Sibboleth,' for he could not pronounce it right" (Judg 12:5–6).

Conventional Star Name/Meaning	Name/Meaning Confirmed or Corrected
Subilon—ear of wheat	confirmed

Enclosed are derivatives and similar words in ancient languages which help clarify the core meaning of this constellation/star name.

Ancient Hebrew/ Aramaic	Phonetic Spelling	Summarized Meaning
Hebrew שִׁבֳּלִים	*shibbol or shibboleth*	ear of grain, branch
Hebrew שִׁבֳּלִים	*šib-bo-lîm*	heads of grain

Ancient Gentile Language	Phonetic Spelling	Summarized Meaning
Assyrian	*si 'bo lit*	an ear, blade (of wheat)
Akkadian	*subiltu*	an ear/spike blade (of barley, corn, wheat) constellation: Virgo, Spica
Arabic	*suba*	branch, offshoot

Scripture References

Gen 41:5, Judg 12:6, Ruth 2:2, Isa 17:5; Zech 4:12, John 12:24

Zaniah

The star name Zaniah (pronounced zah-NYE-a) is a combination of two Hebrew words: אָזַן/*azan*, meaning "hears," and *iah* from יְהֹוָה/Y^{e}hôvâh. The second syllable "*iah*" in Zaniah is a common abbreviation taken from the tetragrammaton—the Hebrew name of God.

Zaniah is an anglicized form of the name of the Israelite יְאַזַנְיָה/Yaa-zaniah, meaning "heard of Y^{e}hôvâh" (2 Kgs 25:23).

Another derivative of Zaniah is זַן/*zan*, which means "kind, sort," and is derived from זוּן/*zûwn* meaning "nourish" or "feed." Zaniah may be defined as "nourished by Y^{e}hôvâh," and is alluded to in the book of Revelation: "But the woman [Virgo] was given two wings of a great eagle, that she might fly into the wilderness to her place, where she is nourished for a time and times and half a time, from the presence of the serpent" (Rev 12:14).

Note: The woman (see color image), having two wings of a great eagle and a robe draped around her, symbolizes Virgo. The wings of this great eagle represent God's divine deliverance for his people:

"You have seen what I did to the Egyptians, and how I bore you on eagles' wings and brought you to Myself" (Exod 19:4).

A similar pronunciation and meanings are found in the selective Gentile languages (see chart). An example of our need to hear from the Lord is found in Isaiah: "Listen to Me, My people; and give ear [*azan*] to Me, O My nation: For law will proceed from Me, and I will make My justice rest as a light of the people" (Isa 51:4).

An example of the Lord hearing our prayers is found in Ps 80: "Give ear [*azan*], O Shepherd of Israel, You who lead Joseph like a flock; You who dwell between the cherubim, shine forth!" (Ps 80:1).

Thank the Lord for hearing the prayers of his people. He delivers us from evil and lovingly provides for his own: "And this is the confidence we have in Him, that, if we ask any thing according to His will, He hears us: And if we know that He hears us, whatsoever we ask, we know that we have the petitions that we desired of Him" (1 John 5:14–15).

Urania's Mirror—Virgo, by Sidney Hall (https://en.wikipedia.org/wiki/File:Sidney_Hall_-_Urania%27s_Mirror_-_Virgo.jpg).

Conventional Star Name/Meaning	Name/Meaning Included
Zaniah—(not included in E. W. Bullinger's book)	included: *Zaniah*—Yᵉhôvâh hears; also: to nourish, feed

Enclosed are derivatives and similar words in ancient languages which help clarify the core meaning of this constellation/star name.

Ancient Hebrew/ Aramaic	Phonetic Spelling	Summarized Meaning
Hebrew אָזַן	azan	hears
Hebrew הַאֲזִנָּה	ha'·zên·nāh	give ear
Hebrew יְהֹוָה	Yᵉhôvâh	LORD
Hebrew יַאֲזַנְיָה	Yaa-zaniah	heard of Yᵉhôvâh (2 Kgs 25:23)
Hebrew זַן	zan	kind, sort; from זוּן/zûwn: to nourish, feed
Aramaic ܙܢܐ	zen, znā	kind, type

Ancient Greek Language	Phonetic Spelling	Summarized Meaning
Assyrian	Yi ʿ hu: wa	Jehovah
Persian	Yāhū	Jehovah
Arabic	Yahwah	Jehovah
Turkish	Yehova	Jehovah
Tigrinya/Ethiopian	Yehowa and Jehova	Jehovah
Chinese	Yē hé huá	Jehovah

Scripture References

Exod 15:26; Deut 32:1; 2 Kgs 25:23; Pss 17:1, 39:12, 78:1; Ezek 8:11, 11:1, 35:3; Ps 80:1; Isa 28:23, 51:4; 1 John 5:14–15; Rev 12:6

Syrma

The star name Syrma (pronounced SIRM-a) is defined in Latin as "a robe with a train." Zodiac charts show the maiden as having a robe draped around her. This robe depicts covenantal marriage to our Maker. The Hebrew word for robe is מְעִיל/mᵉʿîyl, meaning "covering."

According to the Bible, Israel (God's covenant people) was married to God and God was Israel's husband. At Mount Sinai in Arabia, God's people married the Almighty. This is where God's covenant people said, "I do" to all the Lord commanded. It was at Mount Sinai that Israel initially entered into covenant with the God of Israel (Exod 19:8). God, through the prophet Isaiah, said, "For your Maker is your husband, the Lord of hosts is His name; and your Redeemer is the Holy One of Israel; He is called the God of the whole earth" (Isa 54:5).

The message of Virgo extends to the constellation Cassiopeia, whose imagery represents the enthroned bride of Messiah. Cassiopeia represents born-from-above Jews and Gentiles in new covenant relationship with Yeshua—those who are reconciled to the Father and the Father's house through the Messiah.

The description of the robe covering the Bride, the wife of Messiah, is significant. This robe represents the righteousness of Messiah covering his people. We, his people, must be covered in his righteousness. Righteousness means "right standing (with God)." Only by means of Messiah's righteousness do we have access to the throne of the Almighty. The question we must ask ourselves is: Will we stand before the throne of judgment in our own righteousness or his righteousness?

As the verse in Edward Mote's classic hymn "On Christ the Solid Rock I Stand" states: "When he shall come with trumpet sound, / O may I then in him be found, / dressed in his righteousness alone, / faultless to stand before the throne."[2]

In short, once we receive the free gift of righteousness, the Lord teaches us to live according to his lifestyle. The true gift of righteousness and liberty is realized and experienced when we learn to think, speak, and walk even as Yeshua walked. "Now by this we know that we know Him, if we keep His commandments" (1 John 2:3). "He who practices righteousness is righteous, just as He is righteous" (1 John 3:7).

A classic example of the covenantal robe of righteousness is found in Isa 61:10: "I will greatly rejoice in the Lord, My soul shall be joyful in my God; for He has clothed me with the garments of salvation, He has covered me with the robe [מְעִיל/*meil*] of righteousness, as a bridegroom decks himself with ornaments, and as a bride adorns herself with her jewels" (Isa 61:10).

Thank the Lord for the beautiful depiction of his covering over our lives. May we grow in the faith and knowledge of Messiah. He is returning for a bride who has made herself ready—through covenantal agreement—through his love and his word.

> Let us be glad and rejoice and give Him glory, for the marriage of the Lamb has come, and His Bride has made herself ready.

2. Mote, "On Christ the Solid Rock."

And to her it was granted to be arrayed in fine linen, clean and bright, for the fine linen is the righteous acts of the saints. (Rev 19:7–8)

Conventional Star Name/Meaning	Name/Meaning Included
Syrma—(not included in E. W. Bullinger's book)	included: Syrma—robe, train

Enclosed are derivatives and similar words in ancient languages which help clarify the core meaning of this constellation/star name.

Ancient Hebrew/ Aramaic	Phonetic Spelling	Summarized Meaning
Hebrew מְעִיל	mᵉʿîyl	A robe, in the sense of covering; from מָעַל/mâ ʿal (to act covertly, i.e., treacherously: transgress, commit a trespass).

Ancient Greek Language	Phonetic Spelling	Summarized Meaning
Latin	Syrma	a robe with a train
Greek	Phórema	dress, robe

Scripture References

Isa 61:10; Lev 5:15; Exod 19:8; Isa 54:5; 1 John 2:3, 3:7; Rev 19:7–8

2

The Constellation: Coma

Nineteenth-century drawing of *Dendera Zodiac Child with Mother*. Approximately two thousand years before the birth of the Messiah, the Dendera Zodiac, ancient Egyptian bas-relief art, depicts an image of the star Spica—called Nefer, meaning "beautiful" (child) portraying the infant prince held in his mother's (Virgo) arms.

This image indicates Messiah, the seed of the woman (virgin), will be exalted above the woman (virgin). The woman symbolizes Israel, specifically Miriam, mother of Yeshua.

THE ANCIENT HEBREW WORD כָּמַה/kâmah is phonetically similar to the word *coma*, which means "to pine after; long for." This word is used in Ps 63 where David said to God, "My soul thirsts for You; my flesh longs [כָּמַה/kâmah] for You" (Ps 63:1). Similarly, חֶמְדָּה/*chemdah*, means "the desired, the longed for, goodly, pleasant, precious."

Chemdah is used in Haggai's classic prophecy: "and I will shake all nations, and they shall come to the Desire [*Chemdah*] of All Nations, and I will fill this temple with glory, says the Lord of hosts" (Hag 2:7). The desire of ancient Israel and the Gentile nations was the manifestation of the promised Messiah.

Chemdah/חֶמְדָּה is a derivative of the Hebrew word חָמַד/*châmad*, meaning "to desire, covet, take pleasure in, delight in." An example of *châmad* is found in the book of Psalms: "More to be desired [*châmad*] are they than gold, yea, than much fine gold; sweeter also than honey and the honeycomb" (Ps 19:10).

The constellation Coma is a decan, or part, of Virgo and a major part of its prophetic picture and message:

1. In Sumerian culture, the name Coma was *he-gal-a-a*, meaning "great good son."

2. The Egyptian name for this constellation is *Shes-nu*, meaning "Desired Son."

3. The Egyptian Dendera Zodiac conveys the imagery of the star Spica (in Virgo), called Nefer, meaning "beautiful" (child)—portrayed as the infant prince held in his mother's (Virgo) arms. The image of the child is held higher than that of the mother, indicating the child will be exalted above the mother. The mother represents Israel—specifically Miriam, the mother of Yeshua. Fittingly, the constellation Boötes represents Messiah—the exalted shepherd—and is positioned directly above the constellation Virgo.

4. The Hebrew name for Spica is *Tsemach*—"branch," and *Zera*—"seed."

The book of Revelation explains the birth of the *Desired One*: "She bore a male Child who was to rule all nations with a rod of iron. And her Child was caught up to God and His throne" (Rev 12:5).

Egyptian mythology indicates that the Egyptians did not know who this promised child would be. Evidence suggests, however, that the Egyptians should have known that the *Desired One* would come from the Hebrews.

According to Josephus, the Jewish historian, Abraham delivered the science of astronomy to the Egyptians:

> For whereas the Egyptians were formerly addicted to different customs, and despised one another's sacred and accustomed rites, and were very angry one with another on that account; Abram conferred with each of them, and confuting the reasonings they made use of, every one for their own practices, he demonstrated that such reasonings were vain, and void of truth. Whereupon he was admired by them, in those conferences, as a very wise man, and one of great sagacity when he discoursed on any subject he undertook; and this not only in understanding it, but in persuading other men also to assent to him. He communicated to them Arithmetick; and delivered to them the science of Astronomy. For before Abram came into Egypt, they were unacquainted with those parts of learning: for that science came from the Chaldeans into Egypt; and from thence to the Greeks also.[1]

Coma, *the Desired Son*, encompasses the birth, death, resurrection, and exaltation of Yeshua:

> For unto us a Child is born,
> Unto us a Son is given;
> And the government will be upon His shoulder.
> And His name will be called Wonderful, Counselor, Mighty God,
> Everlasting Father, Prince of Peace.
> Of the increase of His government and peace
> There will be no end,
> Upon the throne of David and over His kingdom,
> To order it and establish it with judgment and justice
> From that time forward, even forever.
> The zeal of the Lord of hosts will perform this. (Isa 9:6–7)
> And Jesus increased in wisdom and stature, and in favor with God and men. (Luke 2:52)
>
> I will make him My firstborn,
> The highest of the kings of the earth. (Ps 89:27)

Yeshua is depicted as "the grain of wheat" in the Virgin's left hand: "Most assuredly, I say to you, unless a grain of wheat falls into the ground and dies, it remains alone; but if it dies, it produces much grain. He who

---

1. Josephus, *Works of Josephus*, "Antiquities of the Jews," 1.166–68.

loves his life will lose it, and he who hates his life in this world will keep it for eternal life. If anyone serves Me, let him follow Me; and where I am, there My servant will be also. If anyone serves Me, him My Father will honor" (John 12:24–26).

Virgo, with the accompanying smaller constellation, Coma, is a picture of the birth of the little Prince Messiah. According to the astronomical configuration on September 11, 3 BC, during the Feast of Trumpets, the Sun was in Virgo, the new Moon was beneath her feet, and Jupiter, Venus, and Mercury (including the nine stars of Leo, the Lion), constituted the twelve stars of her crown. This great celestial sign, as recorded in Revelation chapter 12, points to the birth of the Messiah.

"Now a great sign appeared in heaven: a woman clothed with the sun, with the moon under her feet, and on her head a garland of twelve stars. Then being with child, she cried out in labor and in pain to give birth" (Rev 12:1, 2). This marvelous sign is not merely symbolic. It appeared, literally, in the heavens according to God's predetermined time.

The universe is a celestial clock, precise to the second. This means the Creator calibrated the universe to announce the entrance of the Son of God into the earth. Only God could do this. Man cannot calibrate the universe, nor could he manufacture the miraculous conception of Miriam—the incarnation of the Son of God—into the world at a specific time. "But when the fullness of the time had come, God sent forth His Son, born of a woman, born under the law" (Gal 4:4).

Yeshua entered the earth when civilization was approximately *four* thousand years old—as it converged into the *fifth* millennium. In order for Yeshua, the agent of creation, to enter the earth zone he had to enter through the door of humanity. The Hebrew letter for *four* is the *dalet*. The word *dalet* is traditionally associated with the Hebrew word for "door." In biblical numerology, four denotes "universal" and *five* denotes "grace." God sent forth his Son at a specific time in history to provide salvation to humanity. Yeshua's salvation extends from Adam (whom God clothed in ram's skin) to the last person on Earth who will receive his gracious gift of love and salvation through the Lord Jesus, the Messiah. The gospel of Messiah is for *all* people and languages.

The long-awaited promised Seed of the Woman (Yeshua, the Messiah) came two thousand years ago. He will return the second time in the not-too-distant future. Are you ready?

Please pray this prayer with me:

Dear Yeshua, thank you for becoming a Jewish man
and showing me the way of salvation.
Thank you for dying for me on the cross of Golgotha two thousand years ago.
Thank you for paying my sin debt and providing a full pardon for me.
I receive you into my heart and life, here and now.
Thank you for restoring me to the Father of Life.
I am your child. Teach me to follow you by daily feeding on your word.
Fill me with your precious Holy Spirit.
Lead me to be water baptized.
Thank you, Yeshua, for saving me!
Thank you for the love and grace you have shown me today.
In your precious name I pray, Amen.

Christ was offered once to bear the sins of many. To those who eagerly wait for Him He will appear a second time, apart from sin, for salvation. (Heb 9:28)

My star song for Virgo and Coma is titled "Woman Clothed with the Sun."

Woman Clothed with the Sun

(Rev 12)
By Bruce J. Patterson

Verse 1
See a great sign appear
See a great sign appear
In heaven, in heaven
In heaven, in heaven
A woman clothed with the Sun

A woman clothed with the Sun
And the Moon beneath her feet
And the Moon beneath her feet

Verse 2
Upon her head
Upon her head
Is a crown, is a crown
Is a crown of twelve stars
She gave birth
She gave birth
To a Son, to a Son
The Messiah, Holy One

Chorus
Rejoice, rejoice,
For this woman is clothed with the Sun
Rejoice, rejoice,
For her child who is born is God's Son

Verse 3
And He shall rule
He shall rule
All nations, all nations
With an iron scepter
And He was caught up
He was caught up
Unto God and his throne
Unto God and his throne

Chorus

When we look up and contemplate the constellation Coma, may we have a greater appreciation for this glorious sign and God's signature in the stars.

Constellation Name	Translation
Comae Berenices (Latin)	Berenice's hair
he-gal-a-a (Sumerian)	great good son
Shes-nu (Egyptian)	desired son

Enclosed are derivatives and similar words in ancient languages which help clarify the core meaning of this constellation/star name.

Ancient Hebrew/ Aramaic	Phonetic Spelling	Summarized Meaning
Hebrew כָּמַהּ	kâmah	to pine after; long for
Hebrew חֶמְדָּה	chemdah	desire, goodly, pleasant, precious from the Hebrew word חָמַד/châmad, meaning "to desire, covet, take pleasure in, delight in"

Ancient Gentile Language	Phonetic Spelling	Summarized Meaning
Latin	coma	hair
Greek	komā	hair
Sumerian	kam	desire
Assyrian	khma sa	to grasp
Arabic	ḥ	yearn
Hindi	chaahna	to desire
Persian	hamwā	desire
Sanskrit	kAma	longing
Chinese	kài	to desire
Coptic	coume	Virgo, zodiac sign

Scripture References

Hag 2:7; Ps 63:1; Rev 12:5; Isa 9:6–7; Luke 2:52; Ps 89:27; John 12:24–26

3

The Constellation: Centaurus

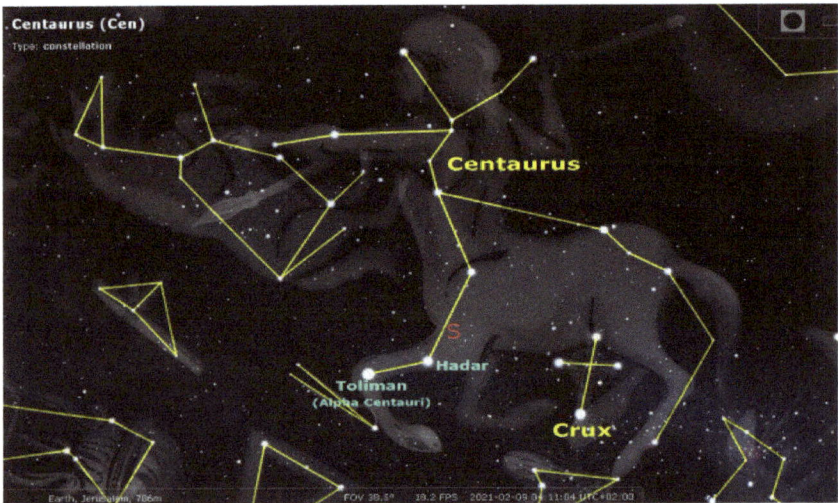

THE NAME *CENTAURUS* IS Latin for "centaur." Since ancient times Centaurus has been one of the fixed signs of the zodiac.

Centaurus is one of the brightest constellations in the southern hemisphere. Positioned between Lupus (the sacrifice) and Argo (the ship), these constellations are alluded to in the ancient book of Job as occupying "the chambers of the south" (Job 9:9).

Centaurus, the Centaur, illustrates the combined characteristics of man and horse—both of which are reflective of the Messiah. The man represents Yeshua (the Son of man), whereas the horse represents the Messiah's governmental administration and carrier of the gospel message.

Centaurus is a visual depiction of the dual nature of the Messiah, who is both God and man. This manner of symbolism is found throughout the Bible. For example, the dual nature of the Messiah is revealed in the ark of the covenant, which was made of acacia wood overlaid with pure gold. The acacia wood represents the humanity of the Messiah. Pure gold represents the deity of the Messiah.

Centaurus is pictorially piercing a sacrifice with a spear. The animal being pierced is known by its Latin name, Lupus, the wolf. The Greeks, however, called this constellation *therion*, meaning "wild animal." The Hebrew word for "wild animal" is בְּהֵמָה/*bᵉhêmâh*, meaning "beast, cattle, animal, livestock (of domestic animals) wild beasts." The plural form of *bᵉhêmâh* is *bᵉhêmot* which includes horses, donkeys, oxen, cows, sheep, goats, and more. Fittingly, a ram is an adult male sheep.

Within the context of these constellations, the symbolic imagery of the ram best fits the narrative. The ram is reflective of the constellation Aries, a picture of the Passover Lamb of God. Medieval Christians associated Centaurus with the sacrifice of Isaac by Abraham where the Lord provided a ram caught in the thicket. The ram offered in place of Isaac was a type of Yeshua, the Lamb of God.

This cosmic imagery of Centaurus is reflective of the Messiah laying down his life as a sacrifice for sinners. Isaiah said: "Yet it pleased the Lord to bruise Him; He has put Him to grief. When You make His soul an offering for sin, He shall see His seed, He shall prolong His days, and the pleasure of the Lord shall prosper in His hand" (Isa 53:10).

No one took his life, but he freely laid it down (John 10:18). In fact, Crux, the Southern Cross, typifying the crucifixion, is highlighted within this celestial scene.

Centaurus and the piercing of the ram found its prophetic fulfillment at the cross of Calvary. After Yeshua died on the cross, the Roman soldier pierced his side with a spear:

> Then the soldiers came and broke the legs of the first and of the other who was crucified with Him. But when they came to Jesus and saw that He was already dead, they did not break His legs. But one of the soldiers pierced His side with a spear, and immediately blood and water came out. And he who has seen has testified, and his testimony is true; and he knows that he is telling the truth, so that you may believe. For these things

were done that the Scripture should be fulfilled, "Not one of His bones shall be broken." And again another Scripture says, "They shall look on Him whom they pierced." (John 19:32–37; Ps 22:16; Zech 12:10; Isa 53:5)

The Roman centurion who witnessed the crucifixion said: "Truly this was the Son of God!" (Matt 27:54). Centaurus carries the eternal truth that "Yeshua is the Lamb slain before the foundation of the world" (Rev 13:8).

Yeshua is both Shepherd and Lamb, Priest and Sacrifice, Creator and Redeemer. He made the tree upon which he would be nailed.

The Bible teaches the necessity of the cross. Praise God for his everlasting sacrificial love for you and me: "God was in Christ reconciling the world to Himself, not imputing their trespasses to them, and has committed to us the word of reconciliation" (2 Cor 5:19).

The Akkadian name for this constellation is ḥabaṣīrānu (ḥaba ṣīr ānu). It is a combination of three ancient words: (1) Hebrew: hab-ba, "he who comes," (2) Hebrew/Sumerian: sar, "prince," and (3) the Sumerian: anu, "heaven, sky." The combination of these three ancient words means "the coming prince of heaven." The selective language chart provides a breakdown of these words.

The true meaning of ḥabaṣīrānu could only be fulfilled by one prince, "the Prince of Peace, "the Prince of Life," Yeshua Hamashiach.

One of the traditional Hebrew names for the constellation Centaurus is *Asmeath*. *Asmeath* comes from the Hebrew אַשְׁמָה/ *'ashmâh*, meaning "guiltiness, a fault, the presentation of a sin-offering, the offering of a victim for guilt or trespass."

This word is used several times in reference to the old covenant sacrificial system, which found its fulfillment in Yeshua, the Lamb of God who takes away the sin of the world.

The Hebrew Bible gives an example of the word אַשְׁמָה/*ashmâh*:

> If a person sins, and commits any of these things which are forbidden to be done by the commandments of the Lord, though he does not know it, yet he is guilty [אָשֵׁם/*asham*] and shall bear his iniquity. And he shall bring to the priest a ram without blemish from the flock, with your valuation, as a trespass offering [אָשָׁם/*asham*]. So the priest shall make atonement for him regarding his ignorance in which he erred and did not know it, and it shall be forgiven him. It is a trespass offering [אָשָׁם/

asham]; he has certainly trespassed [אָשֵׁם/asham] against the Lord. (Lev 5:17–19)

Another example of the use of אַשְׁמָה/ashmâh, appears in Isaiah's classic chapter, Isa 53, foretelling the glorious gospel of God's sacrificial love for you and me: "Yet it pleased the Lord to bruise Him; He has put Him to grief. When You make His soul an offering [אָשָׁם/asham] for sin, He shall see His seed, He shall prolong His days, and the pleasure of the Lord shall prosper in His hand" (Isa 53:10).

Another sobering example of the use of אַשְׁמָה/ashmâh is in Isaiah chapter 24: "The earth is also defiled under its inhabitants, because they have transgressed the laws, changed the ordinance, broken the everlasting covenant. Therefore the curse has devoured the earth, And those who dwell in it are desolate [אָשֵׁם/asham]. Therefore the inhabitants of the earth are burned, and few men are left" (Isa 24:5–6).

A derivative of ashmâh is used in the Hebrew translation of the Greek New Testament of Matthew's Gospel: "See! Your house is left to you desolate [שָׁמֵם/shamem]" (Matt 23:38).

Isaiah describes the fearful reality of impending judgment coming upon the earth just before the return of Yeshua. Yeshua reminds us of the startling world condition before his return: "And unless those days were shortened, no flesh would be saved; but for the elect's sake those days will be shortened" (Matt 24:22).

The Greek word for אָשָׁם/asham is ἁμαρτία/hamartía, meaning "a sin, an offering for our sin, to be sin—or a sin-offering."

"For He made Him who knew no sin [ἁμαρτία/hamartia] to be sin [ἁμαρτία/hamartia] for us, that we might become the righteousness of God in Him" (2 Cor 5:21).

Yeshua did not become a sinner on the cross. He became the sin-offering for our sin. For example, the ancient brazen serpent Moses placed on a pole served as a type of the sacrificial death of the Messiah. Brass biblically typifies judgment. The brazen serpent was a symbol of the curse. The Messiah was made a curse for us on the cross and thereby "swallowed" the curse. Just as there was no poison in the serpent of brass, there was no sin in the Messiah. God acted for our sake. This means God acted—out of his love for us—to remove the separation between him and you and me. To accomplish this, God made the Messiah—who remained sinless during his life on Earth—to become our sin, the sin we inherited through Adam, the corporate head of the human race. Yeshua died in our place, bearing our guilt, removing the obstacle between us and God. Instead of "being sin," those who come to God through faith in the Messiah are given credit for Yeshua's righteous, sinless life. We become the righteousness of God in Christ; God's righteousness, "right standing with God." Through Yeshua, God's Son, we are reconciled to the Father of life so that we may experience a new and vital relationship with the living God.

Another traditional Hebrew name for the constellation Centaurus is *Bezeh*, from בָּזָה/*bâzâh*, meaning "despised."

An example of the use of this word is found in Isaiah's classic chapter, foretelling how the Messiah would be despised בָּזָה/*bâzâh* and rejected by men:

> He is despised [הִזְב *bâzâh*] and rejected by men,
> A Man of sorrows and acquainted with grief.
> And we hid, as it were, our faces from Him;
> He was despised [הִזְב *bâzâh*], and we did not esteem Him.
> Surely He has borne our griefs
> And carried our sorrows;
> Yet we esteemed Him stricken,
> Smitten by God, and afflicted.
> But He was wounded for our transgressions,
> He was bruised for our iniquities;
> The chastisement for our peace was upon Him,
> And by His stripes we are healed. (Isa 53:3–5)

Yeshua reminds his followers that they will likewise be despised: "If the world hates you, you know that it hated Me before it hated you" (John 15:18). "Do not marvel, my brethren, if the world hates you" (1 John 3:13).

Let us remember these verses that we may keep our eyes on Yeshua.

In consideration of the Hebrew name for this constellation, *Bezeh*, "despised," we must beware of religious leaders who despise the true Messiah of Israel and the followers of the Messiah. We must also beware of those who despise prophecies concerning him. The scripture says: "Do not despise prophecies" (2 Thess 5:20).

Thank the Lord whose perfect sacrifice provides "right-standing" before the throne of God. Now we can experience his shalom, healing and wholeness in our spirit, soul, and body. Our ultimate healing occurs when the Messiah returns at the resurrection of the righteous and we put on immortality.

The constellations Centaurus, Lupus, and Crux could be seen descending in the southern sky of Jerusalem the night following the crucifixion. These southern constellations echo the universal reality that Christ laid down his life for us and sin must be judged. One bears one's own guilt or receives Yeshua's sacrifice—the sacrifice he made on their behalf: "Who Himself bore our sins in His own body on the tree, that we, having died to sins, might live for righteousness—by whose stripes you were healed" (1 Pet 2:24).

The heavens truly declare the glory of God, and the glory of God is Yeshua (Ps 19:1)!

My star song for Centaurus is titled "The Heretofore, the Hereafter."

The Heretofore, the Hereafter

(Isa 53; Mic 5:2; Ps 90:2; Rev 1)
By Bruce J. Patterson

The constellation of Centaurus (the Centaur). (Image by Amy Manson, 1895; from *The Gospel in the Stars* by Joseph A. Seiss.)

Verse 1
He is the heretofore, he is the hereafter
Christ, the eternal one, salvation's great captain
Conquering horseman, pierced he died, was buried
But he rose again

Verse 2
He is the Alpha, he is the Omega
The beginning and the end, though despised by men
Conquering horseman, he leads heaven's army
To his triumphant return

Verse 3
He is the Aleph, he is the Tav
The first and the last
The future and past
Conquering horseman, he leads heaven's army
To his triumphant return

Constellation Name	Translation
Centaurus (Latin)	centaur
Asmeath (Hebrew)	sin offering
Bezeh (Hebrew)	despised, contempt
ḫabaṣīrānu (Akkadian)	the coming of the prince of heaven
bàn rén mǎ zuò (Chinese)	the centaur constellation

Enclosed are derivatives and similar words in ancient languages which help clarify the core meaning of this constellation/star name.

Ancient Hebrew/ Aramaic	Phonetic Spelling	Summarized Meaning
Hebrew אַשְׁמָה	ashmâh	guiltiness, a fault, the presentation of a sin-offering—offend, sin (cause of) trespass(-ing, offering)
Hebrew אָשָׁם	asham	sin offering
Hebrew	ă·šā·mām	guilt offering
Hebrew בָּזָה	bâzâh	to despise, hold in contempt, disdain
Aramaic	bwzh	contempt

Ancient Hebrew/ Aramaic	Phonetic Spelling	Summarized Meaning
Hebrew הַבָּא	hab-bā	he who comes
Hebrew שַׂר	sar	prince, ruler

Ancient Gentile Language	Phonetic Spelling	Summarized Meaning
Latin	Centaurus	centaur
Arabic	ḥarām	sin, transgression
Assyrian	bu za kha	rejected, scorn, contempt, disdain, mocking reject
Sumerian	an	sky, heaven
Sumerian	ANU	the supreme God, the heavenly one
Akkadian	agû	sky, climate, wave, flood; wrath, anger
Sumerian	sir	to shine brightly (cf. *sír*) (Akkadian loan from *šaraapu* [m], "to burn").
Assyrian	bâ'u	come
Akkadian	bau	come
Phoenician	bw	come
Arabic	bā'a	return
Chinese	bō	issue, dispatch, send out
Assyrian	sar	chief
Ancient Egyptian	ser	official
Hindi	sardaar	prince
Sumerian	sar	king of
Sumerian	a-ba	who

Scripture References

John 10:18; Lev 4:3, 6:5; Isa 53:3; Song 8:7; Zech 4:10; Num 18:9; John 19:32–37, 15:18; Ps 22:16; Zech 12:10; Isa 53:5

The Stars

Toliman

TOLIMAN IS THE ANCIENT name for the modern star Alpha Centauri. Toliman is the closest star to Earth. This star is actually three stars which overlap in the Alpha Centauri system, appearing to the naked eye as a single bright star. Yet, these stars are an average of 4.3 light-years from Earth.

The Alpha Centauri system was discovered and named in 1915. My good friend and one of my editors, Tim Mitchell, pointed out to me the three stars comprising Alpha Centauri could represent the Holy Trinity (three-in-one)! And I agree. The eternal God is revealed as Father, Son, and Holy Spirit. "Hear, O Israel: The Lord our God, the Lord is one!" (Deut 6:4). "*There is* one Lord, one faith, one baptism. One God and Father of all; who is above all, and through all, and in you, all" (Eph 4:5–6).

Toliman is from the Hebrew עוֹלָם/*olam*, meaning "forever, everlasting, eternal, ancient, always." The prefix ת (*tav*) means "you shall." *Tav* is the last letter of the Hebrew alphabet, meaning "mark, signature, sign, or seal." It is the symbol of truth, perfection, and completion. The *aleph* is the first letter of the Hebrew alphabet. The word *aleph* means "beginning." Indeed, Yeshua is the *Aleph* and *Tav*—the beginning and the end—and everything in between. Yeshua is the "*et*" found in the first verse of the book of Genesis. "*B-re'shiyt bara Elohim 'et hashamayim v'et ha'aretz*," translated "In the beginning God created the heavens and the earth" (Gen 1:1). This two-letter word *et* (אֵת) is the *aleph* and tav. The Hebrew word for truth is אֱמֶת/ *'emet* (*amet*), composed of the *aleph* and *tav*, with the מ "m" in between. Therefore, Yeshua is the embodiment of eternal truth. This is why the Scriptures reveal the Father as the architect of creation and Yeshua as the agent of creation: "God who created all things through Jesus Christ" (Eph 3:9). In fact, Jesus's Hebrew name, Yeshua, means "Yᵉ ᵉhôvâh is salvation."

A timeless example of the use of *olam* is found in the messianic prophecy Micah gave regarding the eternal deity of the Messiah who would be born in Bethlehem. "But you, Bethlehem Ephrathah, though you are little among the thousands of Judah, yet out of you shall come forth to Me the One to be Ruler in Israel, whose goings forth are from of old, from everlasting [עוֹלָם/*olam*]" (Mic 5:2).

Olam is also used in the classic messianic prophecy regarding the reigning Messiah found in Ps 45: "But to the Son He says: 'Your throne, O God, is forever [עוֹלָם/*olam*] and ever; a scepter of righteousness is the scepter of Your kingdom'" (Ps 45:6 and Heb 1:8).

The book of Hebrews quotes Ps 45:6 ascribing the word עוֹלָם/*olam* specifically to the Messiah. The eternal Father declares that Yeshua is the eternal enthroned God. The ancient Jewish translators considered this passage of Scripture positively messianic, referring to the Messiah.

A similar pronunciation and meaning of *olam* is found in the selective Gentile languages (see chart).

God has given us the double-edged sword of revelation (astronomy and the Bible); both of which declare that Yeshua is the eternal עוֹלָם/*olam* King of kings and LORD of lords.

Conventional Star Name/Meaning	Name/Meaning Confirmed or Corrected
Toliman—the heretofore and the hereafter	confirmed: forever, everlasting

Enclosed are derivatives and similar words in ancient languages which help clarify the core meaning of this constellation/star name.

Ancient Hebrew/ Aramaic	Phonetic Spelling	Summarized Meanings
Hebrew ת	tav	mark, sign, signature signal, monument
Hebrew עוֹלָם	olam	forever, everlasting, eternal, ancient, always
Hebrew הָעוֹלָם	hā·ʿō·w·lām	from everlasting
Aramaic ܥܠܡܐ	aLMaA	age, eternity, world
Semitic Roots	-l-m	from the root ع ل م (ʿ-l-m), related to the verb عَلِمَ (ʿalima, "to know"). Related to Aramaic (ʿāləmā), Hebrew עוֹלָם (ōlām)

Ancient Gentile Language	Phonetic Spelling	Summarized Meanings
Arabic	ṭālamā	long, provided
Arabic	Ālam	world
Assyrian	a:l ma	world
Assyrian	a:l mi:n	eternity
Persian	alam	world
Turkish	alem	world
Hindi	alam	world
Tamil	akilam	world
Chinese	yong	forever

Scripture References

Mic 5:2; Isa 44:7; Deut 32:7; Pss 10:16, 41:13, 103:17; Isa 55:3, 56:5; Dan 12:2; Deut 6:4; Eph 4:5, 3:9

Hadar

The star name Hadar is Hebrew (הָדָר/*hâdâr*) for "magnificence, splendor, beauty, comeliness, excellency, glorious, glory, goodly, honor, majesty."

A similar pronunciation and meaning is found in the selective Gentile languages (see chart).

The Hebrew word *hadar* is found several places in the Bible. The constellations Centaurus, Sagittarius and Pegasus are a reflection of psalm 45. The word הָדָר/*hâdâr* is employed by the sons of Korah regarding the eternal majesty of the Messiah: "Gird Your sword upon Your thigh, O Mighty One, with Your glory and Your majesty [הָדָר/*hâdâr*]. And in Your majesty [הָדָר/*hâdâr*] ride prosperously because of truth and meekness and righteousness; and Your right hand shall teach You awesome things" (Ps 45:3, 4).

Isaiah uses the word הָדָר/*hâdâr* in his prophecy regarding the crucifixion of the Messiah:

> Who has believed our report?
> And to whom has the arm of the Lord been revealed?
> For He shall grow up before Him as a tender plant,
> And as a root out of dry ground.
> He has no form or comeliness [הִדְרָה/*hâdâr*];
> And when we see Him,
> There is no beauty that we should desire Him.
> He is despised and rejected by men,
> A Man of sorrows and acquainted with grief.
> And we hid, as it were, our faces from Him;
> He was despised, and we did not esteem Him. (Isa 53:1–3)

Isaiah foretells the coming of the eternal Son of God, the God man, despised and rejected by man, fulfilling the prophetic message of Centaurus. Isaiah 52:14 states: "He was marred more than any man."

In the eyes of man, there was no comeliness הָדָר/*hâdâr* in the Messiah who experienced God's wrath poured out upon himself for you and me on the cross. Because he was beaten beyond recognition and nailed to a Roman cross as a common criminal, the Jewish leaders did not understand the paradox of הָדָר/*hâdâr*. They did not understand how God's perfect majestic Messiah could undergo such fate. Yet many came to realize that the cross is God's greatest expression of הָדָר/*hâdâr* to Israel and the world.

It can be difficult to contemplate and comprehend these things. Yet, with God, a little faith can go far.

John 3:16 could be called "the hub of the Bible." John 3:16 is the scripture upon which the word of God is based. In it we see the beautiful (*hâdâr*) of God's mercy and justice. "For God so loved the world that He gave His only begotten Son, that whoever believes in Him should not perish but have everlasting life" (John 3:16).

May you be encouraged to allow the God of heaven to open your eyes to see the beautiful *hâdâr* of the Messiah with the eyes of faith, as is exemplified by this star name.

> "You have put all things in subjection under his feet." For in that He put all in subjection under him, He left nothing that is not put under him. But now we do not yet see all things put under him. But we see Jesus, who was made a little lower than the angels, for the suffering of death crowned with glory and honor [הָדָר/*hâdâr*], that He, by the grace of God, might taste death for everyone. (Heb 2:8–9 and Ps 8:5)

Glory to Yeshua, great is his *hâdâr!* Amen.

⸎

Conventional Star Name/Meaning	Name/Meaning Confirmed or Corrected
Hadar—(not included in E. W. Bullinger's book)	included: *Hadar*—majesty

Enclosed are derivatives and similar words in ancient languages which help clarify the core meaning of this constellation/star name.

Ancient Hebrew/ Aramaic	Phonetic Spelling	Summarized Meaning
Hebrew הָדָר	*hadar*	majesty, glory
Aramaic ܗܕܪܐ	*hadar*	ornament, embellishment, honor
Aramaic	*hdr*	to be glorious
Aramaic	*hdr, hdar*	glory

Ancient Gentile Language	Phonetic Spelling	Summarized Meaning
Arabic	*zāda*	to increase

Scripture References

Lev 19:32, 23:40; Pss 8:5, 45:4; Isa 35:2, 63:1

4

The Constellation: Boötes

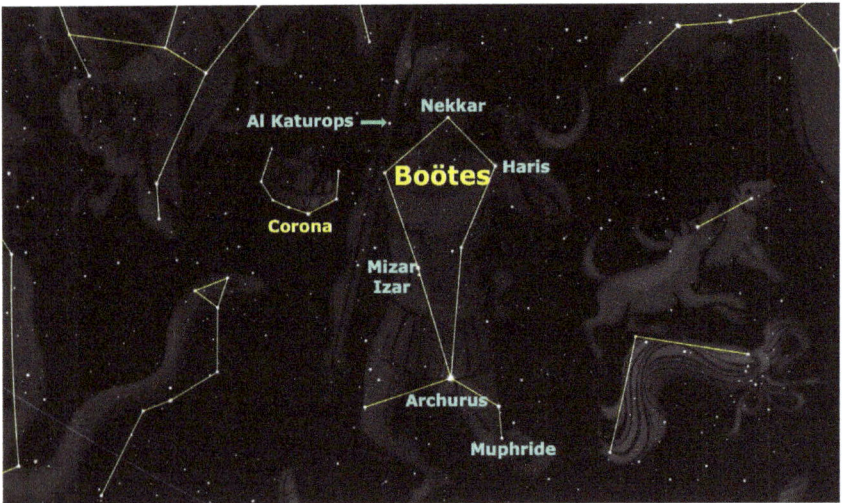

THE WORD BOÖTES IS a combination of two Greek words: *bódi,* meaning "bull, cow, ox," and *óthēsē,* meaning "driver, or to motivate." When combined, these definitions describe Boötes as an "ox-driver or herdsman." From antiquity, Boötes is recognized as the "Herdsman and Shepherd," and is a fixed sign of the zodiac.

The Hebrew name for Boötes is likely *Ayish or Âsh* (from the book of Job).

Ayish or Âsh means "to lend aid, come to help." The primitive root word means "to hasten, assemble" (Job 9:9 and 38:32).

Arabic names for Boötes include Nekkar, "the prodder," and Al Ḥāris al Samā,' "the keeper of heaven," both of which are titled stars in Boötes. In the Chinese language, Boötes is known as *mù fū zuò*, "the shepherd constellation." In the Persian language the name of Boötes is Rāmiḥ, "the one who is armed; who pierces with a spear." To the Chaldeans Boötes was named Papsukal, "the guardian messenger."

As documented by R. Hinckley Allen in his book *Star Names and Their Meanings*, renowned Assyriologists George Smith and the Reverend Archibald Henry Sayce have said that, on the Euphrates River, Boötes is recognized as "the shepherd of the heavenly flock" or "the shepherd of the life of heaven."[1]

The brightest star in the constellation Boötes is Arcturus—the fourth brightest star in the night sky.

Boötes is strategically placed in the northern sky between Corona, the seven-star crown, and Ursa Major, the seven-star constellation, known as the Greater Sheepfold.

Boötes represents the Lord Messiah Yeshua, the Shepherd and Guardian of our souls, guiding and providing for his people as victor, leader, shepherd, and priest.

The Hebrew word *Âsh* can be likened to the ministry of Yeshua, who is (figuratively) assembling his flocks and herds (or "sons") into the heavenly corrals as illustrated by Ursa Major, and Ursa Minor, the Major and Lesser Sheepfolds.

In the book of Job—the oldest book in the Bible—God asks Job the rhetorical question: "Canst thou bring forth Mazzaroth in his season? or canst thou guide [נָחָה/*nachah*] Arcturus [עַיִשׁ/ '*Ayish*, or עָשׁ/ '*Âsh*] with his sons [בֵּן/*bên*]?" (Job 38:32 KJV).

The Hebrew word for *Mazzaroth* refers to the constellations. There are twelve major signs of the Mazzaroth/zodiac and their thirty-six as-sociated "decans" ("parts," in Greek) constellations. This word, מַזָּרָה/ *mazzârâh*, is from נָזַר/*nâzar*, meaning "to dedicate, consecrate, separate," and pertains to a Nazarite who is separated unto God—as were Samuel and Samson—indicating the Mazzaroth is consecrated by God.

The word "guide" (Job 38:32) is נָחָה/*nachah*, meaning "to lead, guide."

The word "Arcturus" (Job 38:32) is עַיִשׁ/ '*Ayish*, or עָשׁ/ '*Âsh*, from עוּשׁ/ '*ûwsh, oosh*; "a primitive root; to hasten; to assemble self."

---

1. Allen, *Star Names*, 101.

The word "sons" (Job 38:32), בֵּן/*bên*, indicates "son," and in the broader sense refers, metaphorically, to sheep, goats, and cattle.

Boötes represents the exalted, heavenly Shepherd and Priest. Boötes is positioned directly above Virgo, the Virgin. Virgo represents Israel and specifically Miriam, the mother of Yeshua. Interestingly, the ancient Egyptian Dendera Zodiac shows an image (see chapter 2, "Coma") of the star Spica—called Nefer, meaning "beautiful" (child). Spica depicts the infant prince Messiah being held in his mother's (Virgo) arms. The head of the son is positioned higher than that of his mother, illustrating that the son will be exalted above the mother. Boötes symbolizes the exalted destiny of this promised son, the true, illustrious Seed of the Woman.

Celestially, Boötes corresponds to Yeshua's statement, "I am the door of the sheep," as illustrated in John chapter 10. In the heavens, Boötes is standing before Ursa Major, the Greater Sheepfold, indicating that Yeshua is the door of the sheep. Boötes represents the humble Shepherd who laid down his life for his sheep.

The constellation Boötes reveals and fulfills the gospel message in several ways. For God to restore humanity to himself, he became a man to pay the penalty for sin—on our behalf. This Good Shepherd became the "Lamb of God who takes away the sins of the world" (John 1:29).

John 3:16 is the focal point of the Bible—and of the heavens: "For God so loved the world that He gave His only begotten Son, that whoever believes in Him should not perish but have everlasting life" (John 3:16). Everything else is commentary on how to live in the love of God.

A reoccurring theme in the heavens is that of the Shepherd, flocks, and herds revealed through these shepherd-related constellations: Boötes the Herdsman/Shepherd/Priest; Ursa Major, the Greater Sheepfold; Ursa Minor, the Lesser Sheepfold; Cancer, the restful sheep corral; Cepheus, the King/Shepherd; and Auriga, the loving and tender Shepherd. The constellations Hercules and Ophiuchus even have shepherd-like qualities and relate to the Bible in unique and wondrous ways. Boötes is particularly significant as a picture of Yeshua, the Shepherd of the sheep, who laid down his life to redeem his people to God.

Belonging and acceptance are two of the deepest needs of humanity. Yet, only in Yeshua, the Messiah, will we find our true belonging and acceptance—as illustrated through the great Shepherd in the sky. No one can live spiritually without being connected to the Father and the Father's house. "Jesus said to him, 'I am the way, the truth, and the life. No one comes to the Father except through Me'" (John 14:6).

I am proposing that Boötes is the prophetic pictogram corresponding to the apostle John's description of Yeshua, the Messiah, in the first chapter of the book of Revelation. Here, he is dressed in high priestly garments walking amid the seven golden lampstands: "Then I turned to see the voice that spoke with me. And having turned I saw seven golden lampstands, and in the midst of the seven lampstands One like the Son of Man, clothed with a garment down to the feet and girded about the chest with a golden band" (Rev 1:12, 13).

Boötes represents the present-day ministry of Yeshua, as High Priest over his house. The seven golden lampstands may correspond to the seven stars of the constellation Ursa Major (the Greater Sheepfold): "These things, says He, who holds the seven stars in His right hand, who walks in the midst of the seven golden lampstands" (Rev 2:1).

To "walk" implies lifestyle, guiding, instructing, and when necessary, prodding with his goad (see my section on the star Nekkar). "Walking among the seven golden lampstands" prophetically declares Yeshua as High Priest, chief Shepherd, instructor, and teacher of the church. He is the Rabbi and Mighty Counselor instructing and leading us in the ways of life.

The book of Revelation reveals the mystery of the seven stars located at the right hand of the Messiah: "Write the things which you have seen, and the things which are, and the things which will take place after this. The mystery of the seven stars which you saw in My right hand, and the seven golden lampstands: The seven stars are the angels of the seven churches, and the seven lampstands which you saw are the seven churches" (Rev 1:19, 20).

"To the angel of the church of Ephesus write, 'These things, says He, who holds the seven stars in His right hand, who walks in the midst of the seven golden lampstands" (Rev 2:1).

Understanding the associated constellation to which John is referring, "having seven stars [at] the right hand of Yeshua," is to consider the position of Boötes in the sky. This begs the question: Could the mystery of the seven stars correspond to the constellation Corona? Yes!

The constellation Corona is composed of seven stars in the shape of a celestial crown. The crown represents dominion lost and dominion found. The dominion Adam lost in the fall is recovered and restored through Yeshua, Israel's Messiah. Directly to the right of Boötes, the seven-star crown of Corona corresponds with Rev 2:1, "These things says

He who holds the seven stars in His right hand, who walks among the seven golden lampstands." The Greek word for "holds" is χρατέω/*kratéō*, pronounced krat-eh'-o, which means "to be strong, to rule, to place under one's grasp." This is precisely what the Messiah accomplished when he recovered and secured Adam's fallen crown by his death, burial, and resurrection. He secured and provided salvation for the whole world.

Note: "The seven stars *in* His right hand," the Greek word "in" or ἐν, pronounced *en*, means "in," as in "realm (sphere) of, at, on the right side." The Greek word for "right-hand" is δεξιός, ά, όν/*dexios*, pronounced dex-ee-os, meaning "on the right hand or right side." The Salkinson-Ginsburg Hebrew New Testament translation of Rev 2:1 (רָאִיתָ בִּימִינִי) also renders "on the right." This passage of Scripture can clearly be translated as "on the right" or "at the right side," and fits the celestial imagery and description of Corona positioned at the right hand of Boötes.

Amazingly, Corona is also located at the right side of the constellation Hercules (*Gibbôr* in Hebrew). In fact, Corona is positioned between Boötes (representing Messiah the Shepherd/Priest), and Hercules (representing Messiah, the Mighty Warrior.) Yeshua, our Mighty Warrior, secured the crown of glory. This crown is worn, symbolically, by the constellation Cepheus, a picture of the Messiah, the King of kings, King of the universe ruling and reigning next to his queen, as portrayed by the constellation Cassiopeia.

In the sky, directly below Corona, the constellation *Serpens*, the serpent, illustrates Satan's attempt to steal and retain the crown of authority God gave Adam and Eve at the beginning of creation. Ophiuchus, "the serpent restrainer," depicts the Messiah preventing the serpent from eternally possessing Adam and Eve's crown.

In essence, the constellation Boötes represents the authority of the Messiah presiding over his kingdom government in his office as Shepherd, Priest, and King.

Each star in Corona corresponds to one of the angels (or pastors) of the seven churches of Asia: "The mystery of the seven stars which you saw in My right hand, and the seven golden lampstands: The seven stars are the angels of the seven churches, and the seven lampstands which you saw are the seven churches" (Rev 1:20).

There are multiple sets of sevens in the book of Revelation, some of which relate to biblical astronomy:

1. Seven is the biblical number of completion and spiritual perfection.

2. The number seven numerically corresponds to *zayin*, the seventh letter of the Hebrew alphabet. *Zayin* represents a weapon or sword in Hebrew and is derived from a root word meaning "sustenance or nourishment." Yeshua, the Messiah, is both our Warrior Protector and Shepherd Provider, as depicted in these warrior/shepherd constellations.

Several constellations are integrated throughout the book of Revelation. These constellations represent real-time occurrences on Earth, some of which are included in *God's Signature in the Stars*. John (Yochanan in Hebrew) wrote the book of Revelation—the words of which were given to John by Yeshua—while in exile on the island of Patmos. Rich in symbolism, the book of Revelation reveals God's truth as directed to the seven churches of Asia in the area which is now modern-day Turkey.

To understand the book of Revelation, one must develop an understanding of the Hebrew scriptures and biblical astronomy. The book of Revelation is a combination of the Bible and astronomy. Astronomy is, symbolically, inscribed upon the pages of the book of Revelation.

"Then I turned to see the voice that spoke with me. And having turned I saw seven golden lampstands, and in the midst of the seven lampstands One like the Son of Man, clothed with a garment down to the feet and girded about the chest with a golden band" (Rev 1:12, 13).

The mystery of the seven golden lampstands corresponds to the seven-branch menorah which was in the holy place of the tabernacle and later the temple. The Hebrew word מְנוֹרָה/*menorah* contains the word אוֹר/*ʾôr*, which means "light." This particular menorah was made of hammered gold. Our lives are like that of a menorah. We are shaped and "hammered," if you will, by the Almighty, that we may reflect his glory as lights to the world. The menorah represents the church (the spiritual Israel of God) and, specifically, the seven churches of Asia.

The menorah may correspond to Ursa Major and the seven stars comprising this constellation. In biblical astronomy, Ursa Major (the Greater Sheepfold) represents the church, "the body of Christ." Metaphorically speaking, the Greater Sheepfold is transformed into a menorah. The "person" standing in the midst of the menorah is a picture of Yeshua, the Messiah, clothed in high priestly garments performing his high priestly duties over his house. Pictorially, this corresponds to the constellation Boötes, the herdsman, the Good Shepherd, guiding and providing for his sheep, the people of God. This imagery may be expanded to include Boötes, metaphorically, walking among the seven stars of Ursa Minor (the Lesser Sheepfold). The Greek word for "standing in the midst" is *mesos*, which means "among, midst, middle, or before."

The constellation Boötes illustrates Yeshua as a Shepherd keeping watch over his flock as he watches over his word to perform it. Furthermore, the Lord spoke to the prophet Jeremiah: "Then the LORD said to me, 'You have seen well, for I am watching over My word to perform it'" (Jer 1:12). In like manner, we, the people of God, are his sheep heeding the voice (word) of our Shepherd. *When we keep God's word in our hearts and mouths, we give our Shepherd/Priest permission to perform his word in our lives as the High Priest of our confession.* "Therefore, holy brethren, partakers of the heavenly calling, consider the Apostle and High Priest of our confession, Christ Jesus" (Heb 3:1).

The word *confession* means "to speak the same as." We are instructed "to speak the same as" the master of the house so we may have the same results.

The book of Revelation cites Yeshua walking amid the candlesticks (the seven churches of Asia) of a menorah. He is watching over his word to perform its outcome on behalf of his new covenant people.

Yeshua said, "And other sheep I have which are not of this fold; them also I must bring, and they will hear My voice; and there will be one flock and one shepherd" (John 10:16).

The "other sheep" refers to redeemed Gentiles, as illustrated in Ursa Major. Ursa Minor may refer to the redeemed Jewish people who receive their Messiah. The two (Jew and Gentile) are becoming one in the Messiah the true heavenly Shepherd of Israel. Yet, as Yeshua watches over his word within the seven churches of Asia, and the modern church today, is he finding our hearts and mouths filled with faith in his word and promises? Each of the seven churches of Asia were given the opportunity to be doers of God's word, so that Yeshua could accomplish great things through them.

In the first three chapters of the book of Revelation, we see Yeshua examining the seven churches of Asia Minor. This examination is a metaphor of the biblical term "to pass under the rod" (Lev 27:32; Jer 33:13; Mic 7:14). The Hebrew Bible presents the example of God counting and examining his sheep and applies it to God counting and examining his people. Metaphorically speaking, only those who "passed the exam" could be brought into the promised land as God's covenant people. Furthermore, the rod may symbolize the cross by which we are redeemed.

The shepherd's rod (שֵׁבֶט/*shebet*) is a two-foot long wooden, club-like rod, likely made of oak. This "club" was the tool under which sheep passed to ensure they were present and accounted for. The process is symbolic of being scrutinized by the word of God. Just as the shepherd runs a rod across his sheep's fleece in order to inspect them for ticks and injuries to the skin beneath its wool, so God watches over his people, to count and evaluate their quality (Heb 4:12, 13).

The first three chapters of the book of Revelation illustrate the Lord's scrutiny and examination of his sheep. To each of the seven churches Yeshua says, "I know your works." As Shepherd, he examines his sheep; as High Priest, he's interceding for his people. He is the complete manifestation of the invisible Shepherd/High Priest walking among the seven golden lampstands.

Interestingly, the celestial shape of the constellation Boötes resembles a first-century Jewish oil lamp. When comparing Boötes to an ancient oil lamp, the position of the star Arcturus pinpoints the place of the flaming wick. There are seven major stars comprising Boötes, in addition to Arcturus.

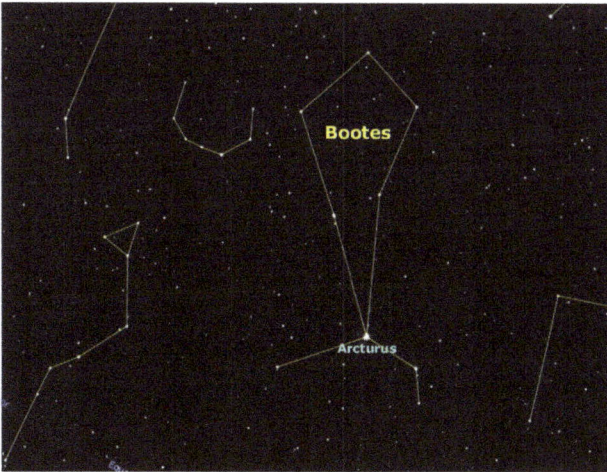

The celestial shape of the constellation Boötes resembles a first-century Jewish oil lamp. (Antique oil lamp image by: Holy Land Market, holylandmarket.com.)

The constellations confirm our confidence in our Creator. May we be like the five wise virgins who keep their lamps burning until Yeshua returns (please read Matt 25:1–13).

When we look up and see the constellation Boötes, may we have a greater appreciation for this glorious sign as God's signature in the stars.

My star song for Boötes is titled "Our Shepherd Knows His Own."

Our Shepherd Knows His Own

(John 10 and Ps 91)
By Bruce J. Patterson

The Lord Is My Shepherd (Image courtesy of Havenlight by Yongsung Kim).

Verse 1
I will rest myself in the faithfulness of God
As I journey this pilgrim way I will rest, I will rest
I will hide myself in the secret place of God
In his shadow I will abide, I will hide, I will hide myself

Chorus
For our loving Shepherd holds on to his own
We're surrounded by his favor
We are safe in his stronghold
For our Shepherd knows his own

<u>**Verse 2**</u>
The enemy is crushed beneath the Lion of Judah
The serpent's seed is vanquished by our King Messiah
The heavenly multitude, the assembled holy seed
Follow the Lamb where he goes; wherever he may lead, he may lead

<u>**Chorus**</u>

Constellation Name	Translation
Boötes (Greek)	herdsman
Nekkar (Arabic)	to prod
Al Ḥāris al Samā' (Arabic)	the keeper of heaven
mù fū zuò (Chinese)	the shepherd constellation
Chaou Yaou, or Teaou (Chinese)	to beckon, excite, or move
rāmiḥ (Persian)	one who is armed or who pierces with a spear
Papsukal (Chaldeans)	the guardian messenger
(on the Euphrates area)	the Shepherd of the heavenly flock, or the Shepherd of the life of heaven

Enclosed are derivatives and similar words in ancient languages which help clarify the core meaning of this constellation/star name.

Ancient Hebrew/ Aramaic	Phonetic Spelling	Summarized Meaning
Hebrew עָיִשׁ	Ayish or Âsh	עָיִשׁ/ 'Ayish, ah'-yish; or עַשׁ/ 'Âsh; the constellation of the Great Bear (perhaps from its migration through the heavens)— Arcturus; from עוּשׁ/ 'ûwsh, oosh; lend aid, come to help, a primitive root; to hasten—assemble self (Job 9:9 and 38:32) Joel 3:11

Ancient Hebrew/ Aramaic	Phonetic Spelling	Summarized Meaning
Hebrew בּוֹא	bôw', bo	to go in, enter, come, go, come in (Ezek 34:13)
Hebrew בּוֹקֵר	bôwqêr (Amos 7:14)	a cattle-tender—herdsman, shepherd, from בָּקַר/ bâqar: to seek, inquire, consider

Ancient Gentile Language	Phonetic Spelling	Summarized Meaning
Greek Βοώτης	Boötes	herdsman
Greek βοσκός	boskós	herdsman, pastor, shepherd, drover
Greek βόδι	bódi	bull, bullock, calf, cattle, cow, ox, steer
Greek ὤθηση	óthēsē	driver, motivate
Greek οδηγητής	odēgētés	driver, mentor
Greek ζευγολάτης	zeugolátēs	plowman
Arabic	baqqār	herdsman, cattle hand
Assyrian	bâ'u	come
Akkadian	bau	come
Phoenician	bw	come
Arabic	bā'a	return
Chinese	bō	issue, dispatch, send out

Scripture References

Job 9:9, 38:32; John 3:16, 14:6; Rev 1:12–13, 1:20, 2:1; Matt 5:14–16; John 5:24–27; Jer 1:12; Heb 3:1, 4:12–13; John 10:16; Ps 21, 121:8; Matt 11:29, 25:1–13; Heb 13:20

The Stars

Arcturus

ARCTURUS IS THE BRIGHTEST star in the constellation Boötes and the fourth brightest star in the night sky.

The word "Arcturus" is a combination of two Greek words, ἄρκτος/ *Árktos,* "bear" and οὖρος/*oûros,* "guardian, watcher." Therefore, Arcturus is said to mean "bear-watcher" or "guardian of the bear." Obviously, this is erroneous in light of the constellation Boötes, the shepherd, who is depicted as guiding flocks and herds. He is not guiding a "bear with cubs." This mythological fairy tale is further falsified by the "tall tail" of the bear. Bears have stubby tails and are predators of sheep. You may recall David, the young shepherd boy, killed both a lion and bear as they attempted to carry a lamb from the flock (1 Sam 17:34–36).

The second Greek word in Arcturus, οὖρος/*oûros,* or "guardian, watcher," accurately conveys the office of the Good Shepherd.

Boötes, symbolically, guards and guides "flocks and herds" associated with its bordering constellations, Ursa Major (Dover, the Greater Sheepfold) and Ursa Minor (Dover, the Lesser Sheepfold). Dover (pronounced "doe-vair") is from the Hebrew word דֹּבֶר/*dover,* "pasture."

Traditionally, these two constellations are known as the Big Dipper and the Little Dipper. The handle of the Big Dipper more accurately represents the "path and door" into the sheepfold of Ursa Major (Dover the Greater Sheepfold). Likewise, the handle of the Little Dipper more accurately represents the "path and door" into the sheepfold of Ursa Minor (Dover, the Lesser Sheepfold).

The interaction of Boötes (likely Ash) to Ursa Major (Dover, the Greater Sheepfold) and Ursa Minor (Dover, the Lesser Sheepfold) is described in the book of Job: "Canst thou bring forth Mazzaroth in his season? or canst thou guide [נָחָה/*nachah*] Arcturus [עַיִשׁ/ *'Ayish*, or עָשׁ/ *'Âsh*] with his sons [בֵּן/*bên*]?" (Job 38:32 KJV).

As stated in the Boötes section, the Hebrew word here for Mazzaroth is מַזָּרֹה/*mazzârâh*, referring to the constellations. There are twelve major signs of the Mazzaroth/zodiac and their thirty-six associated constellations. The word *mazzârâh* is from נָזַר/*nâzar*, meaning "to dedicate, consecrate, separate." It is also used when referring to the Nazarite who is separated to God, as were Samuel and Samson. Accordingly, the Mazzaroth is consecrated by God.

The word *guide* is נָחָה/*nachah*, meaning "to lead, guide."

The word *Arcturus* is עַיִשׁ/ *'Ayish*, or עָשׁ/ *'Âsh*, and means "a constellation" from עוּשׁ/ *'ûwsh, oosh*, "a primitive root, to hasten—assemble self."

The word "sons" (Job 38:32) is בֵּן/*bên*, meaning "son," and in a broader sense can refer, metaphorically, to animals such as sheep, goat, and cattle.

Another example of the Hebrew word נָחָה/*nachah*, "guide," is found in the classic psalm: "The Lord is my shepherd; I shall not want. He makes me to lie down in green pastures; He leads me beside the still waters. He restores my soul; He leads [נָחָה/*nachah*] me in the paths of righteousness For His name's sake" (Ps 23:1–3). Similarly, "So he shepherded them according to the integrity of his heart, and guided [נָחָה/*nachah*] them by the skillfulness of his hands" (Ps 78:72).

There are many examples of the Hebrew word בֵּן/*bên*, meaning "son." Again, in the broader sense, בֵּן/*bên* includes "sheep," as shown in the constellations. Isaiah uses this word in his prophecy regarding the Messiah gathering the sons and daughters of God:

> Fear not, for I am with you;
> I will bring your descendants from the east,
> And gather you from the west;
> I will say to the north, "Give them up!"
> And to the south, "Do not keep them back!"
> Bring My sons [בֵּן/*bên*] from afar,
> And My daughters from the ends of the earth—
> Everyone who is called by My name,
> Whom I have created for My glory;
> I have formed him, yes, I have made him. (Isa 43:5–7)

The sons of God play a central role in creation because all creation is desperately waiting for us to fulfill the assignment our Creator, Redeemer has given us. Biblically speaking, this is the evangelizing of the world and the maturing of the body of Messiah, called the church (the spiritual Israel of God). Known as the "One New Man," it's the mystery of Jew and Gentile becoming one in the Messiah of Israel (Eph 4:11–13).

"For the earnest expectation of the creation eagerly waits for the revealing of the sons of God. For the creation was subjected to futility, not willingly, but because of Him who subjected it in hope; because the creation itself also will be delivered from the bondage of corruption into the glorious liberty of the children of God" (Rom 8:19–21).

Yeshua, our Shepherd, and Priest, enables us to fulfill our calling and purpose. Someday suffering and entropy will end. May we come into agreement with the God of heaven so we may usher in the glorious return of Yeshua, our Messiah. And after the millennial reign of Messiah, we will enter the glory of new heavens and a new earth. Even so, "come, Yeshua."

Finally, and notably, the celestial shape of the constellation Boötes resembles a first-century Jewish oil lamp. When comparing Boötes to an ancient oil lamp, the star, Arcturus, corresponds to the flaming wick.

There are seven major stars contained in Boötes—in addition to Arcturus. Yeshua, the Light of Life, ignites our light as we follow him:

> Then Jesus spoke to them again, saying, "I am the light of the world. He who follows Me shall not walk in darkness, but have the light of life." (John 8:12)

> For with You is the fountain of life; in Your light we see light. (Ps 36:9)

> The spirit of a man is the lamp of the LORD, searching all the inner depths of his heart. (Prov 20:27)

> Yeshua says: "Let your waist be girded and your lamps burning." (Luke 12:35)

Conventional Star Name/Meaning	Name/Meaning Confirmed or Corrected
Arcturus—he comes (Greek)	corrected: bear-watcher to be chief, to lead, to rule
Al Haris al Sama—keeper of heaven (Arabic)	confirmed

Enclosed are derivatives and similar words in ancient languages which help clarify the core meaning of this constellation/star name.

Ancient Hebrew/ Aramaic	Phonetic Spelling	Summarized Meanings
Hebrew עַיִשׁ	*ayish* or *ash* (Job 9:9) "Arcturus"	(from): עוּשׁ/*ush*: (to lend aid, come to help) root: (hasten to assemble)
Hebrew אֵשׁ	*Esh*	fire

Ancient Gentile Language	Phonetic Spelling	Summarized Meanings
Arabic	Al Ḥāris al Samā	Arcturus was once called "the keeper of heaven"
Greek †ἄρκτος	*ark'-tos*	content, enough
Greek ἄρκτος	*arktos*	bear, she-bear
Greek οὖρος	*oûros*	guardian, watcher
Latin	*Arctos, Arctī*	the Great Bear (Ursa Major)
Latin	*tūtor, tūtōris*	a watcher, protector, defender
Latin from Greek	*Arctūrus*	(bear-guard), the brightest star in Boötes
Greek ἄρχω	*archō*	to be first (in political rank or power)—reign (rule) over
Arabic	Al Haris al Sama	keeper of heaven
Arabic	*ḥāris*	keeper
Arabic	*samā*	heaven
Greek οὖρος	*oûros*	guardian, watcher

Scripture References

Job 9:9, 38:32; John 10; Joel 3:11–12; Matt 25:31–46

Nekkar

The star name Nekkar is from the Hebrew נִקַּר/*niq·qar*, meaning "pierces; are pierced" from נָקַר/*nâqar*, meaning "pierced; dig."

A similar pronunciation and meaning is found in the selective Gentile languages (see chart).

The star Nekkar is located at the top of Boötes and may correspond to the point of the heavenly goad he carries. One of the Arabic names for the constellation Boötes is Nekkar, "the prodder." The Chinese called this star Heuen Ko, "the Heavenly Spear."

The shepherd uses a tool called a "goad." The goad causes a נִקַּר/*niq·qar*, or piercing. The goad is a customary farming tool used to spur or guide livestock, usually oxen, as they pull a plough or a cart. It is a type of long pole or stick with a pointed end, also known as a cattle prod.

In the strong hands of a loving master, the ox goad is used to gently prod and guide the animal in the desired direction when plowing fields. When a stubborn ox attempts to kick back against the goad, it causes the animal discomfort. The ox inflicts pain upon itself.

Oxen are created by God to serve and glorify the One who created them. Oxen bend their necks to the yoke to serve their master. The ox goad can be used as a tool of persuasion. Similarly, when man is stubborn and refuses to submit to the revealed will of his Maker, a "goad," of sorts, may serve as an appropriate "persuader."

Even King David experienced the need for God's loving hand to correct and guide him: "Thus, my heart was grieved, and I was vexed in my mind. I was so foolish and ignorant; I was like a beast before You. Nevertheless, I am continually with You; You hold me by my right hand. You will guide me with Your counsel, and afterward, receive me to glory" (Ps 73:21–24).

The book of Ecclesiastes mentions the ox goad: "The words of the wise are like goads, and the words of scholars are like well-driven nails, given by one Shepherd" (Eccl 12:11).

Saul of Tarsus, who became the apostle Paul, is a well-recognized example of one who experienced the figurative "tip of the ox goad." Saul, who ignorantly persecuted messianic (Jewish) believers, could be compared to a stubborn ox. Yet, God chose Saul to be a disciple and apostle of Yeshua. The act of disciplining Saul with the point of the ox goad is symbolically reflected in the star Nekkar.

"As he [Saul] journeyed, he arrived near Damascus. Suddenly a light shone around him from heaven. As he fell to the ground, he heard a voice saying, 'Saul, Saul, why are you persecuting Me?' Saul said, 'Who are You, Lord?' Then the Lord said, 'I am Jesus whom you are persecuting. It is hard for you to kick against the goads' (Acts 9:3–5).

It behooves us to work *with* God rather than against him. Saul got the point—no pun intended. "Although I was formerly a blasphemer, a persecutor, and an insolent man; but I obtained mercy because I did it ignorantly in unbelief" (1 Tim 1:13).

Yeshua is a prime example of the "oxen" metaphor. He likens himself to an ox and considers each disciple a little ox-in-training. "Come to Me, all you who labor and are heavy laden, and I will give you rest. Take My yoke upon you and learn from Me, for I am gentle and lowly in heart, and you will find rest for your souls. For My yoke is easy and My burden is light" (Matt 11:28–30).

The yoke our loving master makes for each of us is custom made. It fits perfectly and does not chafe. When we bend our neck and apply the loving yoke of Yeshua, we find it to be easy and his burden, light.

It is far better to be yoked with Yeshua (the last Adam), than to be yoked with unredeemed mankind—in Adam. To be yoked with Adam places one under the law (under the penalty of the law) called the curse of the law (Torah) which is burdensome and heavy: "Therefore you shall serve your enemies, whom the LORD will send against you, in hunger, in thirst, in nakedness, and in need of everything; and He will put a yoke of iron on your neck until He has destroyed you" (Deut 28:48).

Messiah has legally redeemed us from this heavy yoke of iron: "Christ has redeemed us from the curse of the law, having become a curse for us (for it is written, 'Cursed is everyone who hangs on a tree'), that the blessing of Abraham might come upon the Gentiles in Christ Jesus" (Gal 3:13, 14).

In the Messiah, we are legally redeemed from the heavy burden of sin, sickness, disease, poverty, fear, apprehension of the future, demonic activity, legalism and all enemies.

Praise God, in the Messiah, we find the Lord's commandments to be a delight: "For this is the love of God, that we keep His commandments. And His commandments are not burdensome. For whatever is born of

God overcomes the world. And this is the victory that has overcome the world—our faith" (1 John 5:3, 4).

When we are yoked with Yeshua, there is no need for the pointed end of the goad—the Nekkar effect. As we learn to keep step with Yeshua, likened to the "bigger ox," we realize Yeshua is doing most of the work and we, as the "smaller ox," can enjoy our journey.

The Plowman by Balage Balogh, Archaeology Illustrated

Conventional Star Name/Meaning	Name/Meaning Confirmed or Corrected
Nekkar—the pierced	confirmed

Enclosed are derivatives and similar words in ancient languages which help clarify the core meaning of this constellation/star name.

Ancient Hebrew/ Aramaic	Phonetic Spelling	Summarized Meaning
Hebrew נָקַר	nâqar (naw-kar)	pierced, dig niqqar = pierces
Hebrew נִקַּר	niq·qar	pierces, are pierced
Hebrew נֹקֵד	nôqêd	a spotter (of sheep or cattle), i.e., the owner or tender (who thus marks them)—herdsman, sheep master From נָקֹד/nâqôd, to mark (by puncturing or branding); spotted—speckled
Aramaic ܢܩܒ	NQaB	bore, pierce
Aramaic ܢܩܝܐ	NeQYaA	sheep
Aramaic	nqrh, nqrt	cleft (in the cleft of the flint rock)

Ancient Gentile Language	Phonetic Spelling	Summarized Meaning
Arabic	naqqaba	pierce
Arabic	naqaba	bore
Arabic	naḵaza	to goad
Arabic	nakza	to prod
Akkadian	naqābu	to pierce
Assyrian	nqa ba	bore, to pierce
Assyrian	na:q da	a shepherd
Akkadian	nāqidu	a shepherd, herdsman, rancher (sheep, goats, cows)
Persian	niqād	niqād (pl. of naqad), certain kinds of sheep of a bad sort, a shepherd of the sheep called naqad
Persian	naqqād	certain kinds of sheep of a bad sort
Chinese	Heuen Ko	the heavenly pear

Scripture References

Deut 28:48; Ps 73:21–22; Eccl 12:11; Job 30:17; Isa 51:1; Zech 12:10; Acts 9:3–5; 1 Tim 1:13; Matt 11:28–30; Gal 3:13–14; 1 John 5:3–4

Mizar

The word *mizar* means "little" in Hebrew. The Arabic and Persian definition, however, is more applicable when referring to the constellation Boötes.

In Arabic and Persian, *mizar* means "apron." The Arabic word, *mizar*, a derivative of the word *Izar*, is a secondary name for the star. *Izar* is a short Arabic phrase meaning the "girdle" or "loin cloth."

Boötes is traditionally depicted wearing an apron type garment. A corresponding Hebrew word for the *mizar* (a type of apron) is אֵפוֹד/ *ephod*, meaning "a girdle, apron-like garment." Boötes likely corresponds to the imagery of the Messiah, in the first chapter of the book of Revelation, wearing his high priestly garments—including the ephod (apron).

The ephod, a symbol of service, is an apron-like garment made of fine linen worn over the robe of the high priest. The ephod is embroidered with blue, purple, scarlet, and white—the same colors used in the tabernacle furnishings. These colors could be seen at the door to the outer court, the entrance to the sanctuary, and through the veil (Exod 28:6–14; 39:2–7).

The prophet Samuel was dedicated to God, at birth, by his mother, Hannah. As a child, Samuel wore a linen ephod as he ministered to the Lord (1 Sam 2:18–26).

King David wore a linen ephod when he danced before the Lord as the ark of the covenant was carried into Jerusalem. "Then David danced before the Lord with all his might; and David was wearing a linen ephod" (2 Sam 6:14).

The most significant ephod is worn, figuratively, by Yeshua, the son of David, the reigning Messiah himself:

> Then I turned to see the voice that spoke with me. And having turned I saw seven golden lampstands, and in the midst of the seven lampstands One like the Son of Man, clothed with a garment down to the feet and girded about the chest with a golden band. His head and hair were white like wool, as white as snow, and His eyes like a flame of fire; His feet were like fine brass, as if

refined in a furnace, and His voice as the sound of many waters; He had in His right hand *seven stars*, out of His mouth went a sharp two-edged sword, and His countenance was like the sun shining in its strength. (Rev 1:12–16)

The original Aramaic New Testament uses the word *ephod* in the apostle John's description of the Messiah's heavenly attire: "And in the midst of the menorahs as the likeness of a man, and he wore an ephod* and he was girded around his chest with a golden wrap" (Rev 1:13 OANT).

The word garment in Rev 1:13 is the Greek word ποδήρη/*podérés* used for "ephod" in the Greek Septuagint translation of the Hebrew Bible describing the priestly attire (Exod 28:4). This garment is the same attire worn by the high priest during the tabernacle service.

The first chapter of the book of Revelation is a picture of the immortal glorified Messiah (Boötes) reigning in glory. He is clothed in high priestly attire interceding for you and me. At his right hand is Corona, the seven-star crown, representing authority restored. The dominion Adam lost, Yeshua (the Son of Man, the last Adam) has restored, yes, and even more!

Halleluyah to his name!

Conventional Star Name/Meaning	Name/Meaning Confirmed or Corrected
Mizar—guarding	corrected: apron

Enclosed are derivatives and similar words in ancient languages which help clarify the core meaning of this constellation/star name.

Ancient Hebrew/ Aramaic	Phonetic Spelling	Summarized Meanings
Hebrew אֵפוֹד	*ephod*	a girdle, apron-like garment, a type of apron
Hebrew מִצְעָר	*mits'âr*	small, little, few

Ancient Gentile Language	Phonetic Spelling	Summarized Meanings
Arabic	*mi'zar*	apron
Persian	*mi'zar*	apron
Arabic	*nazīr*	little
Akkadian	*maṭû*	little
Sanskrit	*mita*	little

Scripture References

Exod 28:4; 1 Sam 2:18–26; 2 Sam 6:14; Ps 42:6; Isa 24:6; Luke 12:32; Rev 1:12–16

Muphrid

The star name Muphrid is Arabic for "single." The Semitic root of Muphrid is *prd*, meaning "to separate, single." The corresponding Hebrew word is פָּרַד/*parad*, meaning "separated, to divide."

An example of the use of this Hebrew word is found in Deuteronomy: "When the Highest divided to the nations their inheritance, when He separated [*parad*] the sons of Adam, He set the bounds of the people according to the number of the children of Israel. For the Lord's portion is His people; Jacob is the place of His inheritance" (Deut 32:8).

The God of Israel continues to separate his people (Jew and Gentile) from all the nations of the earth to become his family, one in Messiah. They are his portion as indicated by this star name, Muphrid, in Boötes, the Shepherd.

Another interesting example of this Hebrew word (פָּרַד/*parad*) is found in the book of Genesis: "Then Jacob separated [*parad*] the lambs, and made the flocks face toward the streaked and all the brown in the flock of Laban; but he put his own flocks by themselves and did not put them with Laban's flock" (Gen 30:40).

Thank the Lord for separating his people unto himself as his own special treasure. And one day his people will become jewels in the crown of the Messiah. "Then those who feared the LORD spoke to one another, And the LORD listened and heard them; so a book of remembrance was written before Him for those who fear the LORD and who meditate on

His name. 'They shall be Mine,' says the LORD of hosts, 'On the day that I make them My jewels. And I will spare them as a man spares his own son who serves him'" (Mal 3:16, 17).

Conventional Star Name/Meaning	Name/Meaning Confirmed or Corrected
Muphrid—who separates	confirmed

Enclosed are derivatives and similar words in ancient languages which help clarify the core meaning of this constellation/star name.

Ancient Hebrew/ Aramaic	Phonetic Spelling	Summarized Meaning
Hebrew פָּרַד	*parad*	separated, to divide
Aramaic ܦܪܫ	*P'RaSH*	separate
Aramaic	*prd*	glory to flee; separate

Ancient Gentile Language	Phonetic Spelling	Summarized Meaning
Arabic	*mufrad*	single
Persian	*mufrad*	single, alone, solitary, unique
Arabic	*munfarid*	separate, to divide
Turkish	*münferit*	single, solitary
Semitic Roots	*prd*	West Semitic, to separate. *fardel*, from Arabic *farda*, single piece, pack, bundle, from *farada*, to be(come) separate, segregated, single

Scripture References

Gen 13:11, 30:40; Deut 32:8; 2 Cor 6:17

Al Kalurops

The star name Al Kalurops (pronounced al-ka-LOOR-ops) is the Arabic version of the Greek καλαύροψ/*kalaurops*, meaning "the shepherd's crook."

The Hebrew word for the shepherd's crook or staff is מִשְׁעֶנָה/ *mish-ay-naw*.

A prime example of *mish-ay-naw* is found in this classic psalm: "Your rod [שֵׁבֶט/*shebet*] and Your staff [מִשְׁעֶנָה/*mish-ay-naw*], they comfort me" (Ps 23:4).

Here is a brief description of the rod and staff:

The Shepherd's Rod

The rod (שֵׁבֶט/*shebet*) is a relatively short, heavy, club-like device, whereas the staff (מִשְׁעֶנָה/*mish-ay-naw*) is longer and thinner, with a hook or crook at one end.

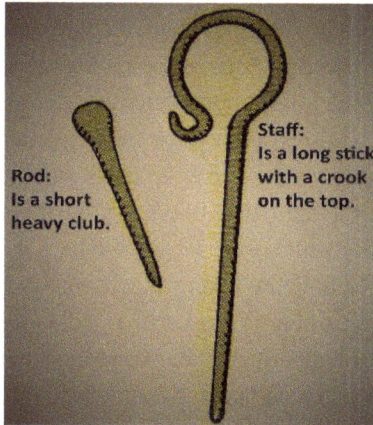

A staff (הֲגְעֶשׁמ/*mish-ay-naw*) was used as a walking staff and defensive weapon. A rod (שׁבֶט/*shebet*) is used for correction, discipline, and guidance. (Drawing by Jim Pinkoski.)

Shepherds customarily carry certain instruments as they tend their flock in the high country. This was especially prevalent in ancient times.

The rod (שֵׁבֶט/*shebet*) was a tool used to protect the shepherd and his flock. It was also used to discipline and correct sheep who wandered from the flock.

The Hebrew Bible presents imagery describing how God examines his people, a method similarly utilized in examining sheep. This sheep examination, called "to pass under the rod," is a figure of speech appropriated from the method a shepherd employs in counting and examining their flock (Lev 27:32; Jer 33:13; Mic 7:14). Today, God uses this "pass under the rod" method as a means by which to determine who qualifies to enter his "spiritual" promised land. Have you received Jesus as your Lord and Shepherd? If so, you will "pass under the rod."

The Shepherd's Staff

The shepherd's staff (מִשְׁעֵנָה/mish-ay-naw) is a long wooden implement with a crook at the top. The staff is used to support the shepherd during his long, hot days of guarding and guiding his sheep. The shepherd staff is also used to reach out and hook sheep in order to draw them to himself. The sheep are, in fact, separated unto the shepherd. This act may relate to the previous star name, Muphrid, or "single," in Arabic, and corresponds to the Hebrew פָּרַד/parad, meaning "separated, to divide."

One of the primary functions of the staff is to guide and protect the sheep from danger. The long crook of the shepherd's staff can lift sheep out of dangerous situations, such as becoming entangled in the brush or falling into rushing water. The shepherd's staff is an instrument of comfort to both the shepherd and his sheep. Furthermore, the staff may symbolize the cross by which we are redeemed.

The Bible tells us how the young shepherd boy David protected his sheep from a lion and a bear (1 Sam 17:34–36). God's people today can know that Yᵉhôvâh's Shepherd, Yeshua, our Messiah, will protect us. We can rest assured he is watching over his people as the Shepherd and Guardian of our souls. This is the message of the star Al Kalurops, in Boötes.

The Lord uses his spiritual staff today to guide us in every area of our lives. Consider these verses in Ps 23: "He leads me beside the still waters . . . He leads me in the paths of righteousness for His name's sake."

Our good Shepherd uses his staff to lead us to places where we can find peace and restoration during the challenges of life. He uses his staff to lead us in paths we should go and helps us make choices in keeping with his will for our lives and families.

Thank the Lord for his rod and staff that corrects and comforts us.

There is not a more beautiful description of our loving Lord than that of the Good Shepherd. The Good Shepherd gives his life for the sheep and provides pasture for his own. The Scripture says, "He calls us by name" (John 10:1–21).

Conventional Star Name/Meaning	Name/Meaning Confirmed or Corrected
Al Kalurops—the branch	corrected: the shepherd's crook

Enclosed are derivatives and similar words in ancient languages which help clarify the core meaning of this constellation/star name.

Ancient Hebrew/ Aramaic	Phonetic Spelling	Summarized Meaning
Aramaic	*krzyl, karzīl, karzīlā*	shepherd's crook; junior shepherd
Hebrew חֹטֵר	*chôṭêr* or *kho'-ter*	branch, twig, rod

Ancient Gentile Language	Phonetic Spelling	Summarized Meaning
Greek καλαύροψ	*kalaurops*	a shepherd's crook, a stick
Assyrian	*qa: ' ṭi: ia*	crook
Greek γκλίτσα	*nklítsa*	shepherd's crook

Scripture References

Ps 23; 1 Sam 17:34–36; John 10

Haris

The star name Haris is Arabic and Persian for "guard, watchman." The corresponding Hebrew word is שָׁמַר/*shamar*, meaning "to keep, watch, preserve." The word *haris* in the Arabic Bible has the same meaning as the Hebrew word *shamar*.

An example of the word *shamar* is used six times in Ps 121:

> He will not allow your foot to be moved;
> He who keeps [רְמַ֫שׁ/*shamar*] you will not slumber.
> Behold, He who keeps [רְמַ֫שׁ/*shamar*] Israel
> Shall neither slumber nor sleep.
> The Lord is your keeper [רְמַ֫שׁ/*shamar*];
> The Lord is your shade at your right hand.
> The sun shall not strike you by day,
> Nor the moon by night.
> The Lord shall preserve [רְמַ֫שׁ/*shamar*] you from all evil;
> He shall preserve [רְמַ֫שׁ/*shamar*] your soul.
> The Lord shall preserve [רְמַ֫שׁ/*shamar*] your going out and your
> coming in
> From this time forth, and even forevermore. (Ps 121:3–8)

While writing this book, COVID-19 took its toll on humanity. God's promises, however, remain true. His love is steadfast. And he keeps and preserves his people who trust in him. He delivers us from every plague and all evil. If a saint passes away from this dreadful disease, yet shall he live. "For if we live, we live to the Lord; and if we die, we die to the Lord. Therefore, whether we live or die, we are the Lord's" (Rom 14:8).

The star name Haris is defined as "guard, watchman," and perfectly describes the constellation Boötes. "Haris" enhances the overall message of Boötes, as "herdsman, pastor, shepherd," and is reminiscent of the Greek word *oûros*, "guardian, watcher," as in the star Arcturus.

Haris also complements the star Al Kalurops. The Arabic translation of the Greek word *kalaurops*, "the shepherd's crook," further emphasizes heaven's Shepherd actively watching over us using his rod and staff for our good.

The star names in Boötes magnify the ministry of Yeshua, as the Great Shepherd of his sheep.

> Now may the God of peace who raised our Lord Jesus from the
> dead, that great Shepherd of the sheep, through the blood of the
> everlasting covenant, make you complete in every good work to

do His will, working in you what is well pleasing in His sight, through Jesus Christ, to whom be glory forever and ever. Amen. (Heb 13:20, 21)

Conventional Star Name/Meaning	Name/Meaning Confirmed or Corrected
Haris—guard, watchman	confirmed

Enclosed are derivatives and similar words in ancient languages which help clarify the core meaning of this constellation/star name.

Ancient Hebrew/ Aramaic	Phonetic Spelling	Summarized Meaning
Hebrew שָׁמַר	*shamar*	to keep, watch, preserve

Ancient Gentile Language	Phonetic Spelling	Summarized Meaning
Persian	*ḥāris*	a keeper; a governor, defender, watchman; a sentinel, guard, a farmer, ploughman; one who gathers
Arabic	*ḥāris*	guard, watchman
Greek χάρις	*chári*	grace, graciousness

Scripture References

Pss 23, 121:3–8; Heb 13:20–21

5

The Constellation: Libra

Honest weights and scales are the Lord's; all the weights in the bag are His work. —Prov 16:11

THE WORD *LIBRA* IS Latin for "scales." Similarly, the Hebrew word for Libra, מֹאזְנַיִם / Moznayim means "scales."

The God of heaven made Libra to illustrate his "scales of justice" and to proclaim his justice and mercy to mankind. He fashioned Libra to illustrate redemption—the price God paid to purchase your life and mine through the death, burial, and resurrection of his only son, Jesus Christ, Yeshua Hamashiach.

America's judicial system uses "scales" as an emblem of justice and fairness. In a court of law, the statue of Lady Justice is blindfolded and holds a set of scales in her hand. Her presence indicates that our system of jurisprudence is legally and morally obligated to weigh, impartially, both sides of evidences presented to the court.

Weighing scales are an essential means by which merchants determine the value of goods. The earliest record of weighing scales dates back four thousand years to the Indus River Valley civilization near present-day Pakistan.

Weighing scales are built with two plates attached to an overhead beam secured to a pivot on a central pole. Measurements are determined by placing an object on one plate and adding stones of a known weight on the other until the weights are equal and the beam horizontal.

The British Museum in London houses one of the oldest Egyptian drawings of weighing scales on papyrus from the Nineteenth Dynasty, New Kingdom, 1275 BCE.[1] According to Egyptian symbolism, these scales present a vertical fulcrum, or "head," of the scales as their view of truth. The Egyptian name for this head is Ma'at, meaning "divine order and ethical life." The word Ma'at is phonetically similar to the ancient Hebrew word אֱמֶת/emet, which means "truth."

This ancient "weighing scales" imagery drawn on Egyptian papyrus illustrates the judgment of an individual who died and is adjudicated through the "feather of truth" principle: The left weighing pan depicts the condition of the "heart" of an individual while being weighed against the feather of truth. The right weighing pan, filled with one "feather of truth," indicates a pure heart worthy of eternal life—if the left weighing pan (heart) balances with the right weighing pan. If the heart did not balance with the feather, then the dead person was sentenced to be devoured by the gnashing teeth of a hideous beast.

Weighing scales symbolize "the measure of a man." Who is the head of the scale of justice? Who is mediating on our behalf? The fulcrum of a weighing scale is designed in the shape of a man standing upright with his arms and hands outstretched, horizontally. Metaphorically, weighing scales represent the Messiah "who, for the joy set before Him endured the cross" (Heb 12:2)—God's eternal "weighing scale."

---

1. British Museum, *Papyrus Weighing Scales*.

There are four major stars in Libra, which contribute to its overall message:

1. Zuben El Genubi, "the scale south" or "the southern scale"

2. Zuben Al Chemali, "the scale north" or "the northern scale"

3. Zuben Akrabi, "scales of the conflict, war"

4. Brachium, the "arm"

(Please note each of the above star names examined in this chapter.)

The Persian name for Libra is Lambadia. Lambadia is a combination of two words: *lam*, meaning, "forgiveness, rest" and *badia*, meaning "eternal." Together, these two words denote "eternal forgiveness and rest." The God of the universe grants eternal forgiveness and rest to the redeemed. The apostle Paul alludes to this concept regarding the saint's eternal weight of glory: "For our light affliction, which is but for a moment, is working for us a far more exceeding and eternal weight of glory" (2 Cor 4:17).

God strategically placed Libra (the scales) between Virgo and Scorpio to illustrate *the Seed of the Woman, Messiah Yeshua, destroying Satan—the perverter of justice and fairness—in order to secure our eternal redemption.*

Libra's weighing scales illustrate the value of a soul and the weighty price of redemption. Yeshua was nailed to a cross in the form of a weighing scale. With his outstretched arms, this "perfect man" paid our sin debt—balancing the scales—redeeming us as his own. Exodus chapter 6 is a picture of the price of redemption Yeshua willingly paid on our behalf. "I will redeem you with an outstretched arm and with great judgments" (Exod 6:6).

Yeshua is the personification of grace and truth which the scales typify (John 1:14). Yeshua, figuratively, stands on the scales of justice on behalf of his redeemed people (Eph 2:8–9). When we invite him into our hearts, he lives his life of righteousness through us so we can learn to appropriate his "balanced" lifestyle (Rom 8:1–4).

Yeshua said, "I am the way, the truth, and the life. No one comes to the Father except through Me" (John 14:6). He made this statement to his twelve Jewish disciples. Today, God's plan of redemption is offered to *all* people. The gospel is to the Jewish people first because God chose Israel to be his special possession (Deut 7:6–11). Thank God, the gospel is for *all* people, nations, tribes and languages: "For I am not ashamed of the gospel of Christ; for it is the power of salvation to the Jew first and also to the Greeks [Gentiles]" (Rom 1:16).

Have you personally accepted and received Yeshua's sacrifice of love on your behalf? Is Yeshua the head of your life and family?

"For there is one God and one Mediator between God and men, the Man Christ Jesus, who gave Himself a ransom for all, to be testified in due time" (1 Tim 2:5, 6).

The constellation Libra is a symbol of redemption. Its starry configuration illustrates the price God required of himself to purchase fallen mankind. Yeshua's purchase of the redeemed includes everyone who receives Christ—God's gracious gift of salvation—from Adam to the last person on Earth. All who are atoned through the blood of the old covenant, who died before Yeshua's work on the cross was completed, are, at last, redeemed by the precious blood of the Lamb of God: "And for this reason, He is the Mediator of the new covenant, by means of death, for the redemption of the transgressions under the first covenant, that those who are called may receive the promise of the eternal inheritance" (Heb 9:15).

All who received God's gracious gift of pardon after the cross, including the saints who are alive at the Lord's return, are redeemed by the precious blood of the Lamb. The blood of the Lamb of God is eternal: "Now may the God of peace who brought up our Lord Jesus from the dead, that great Shepherd of the sheep, through the blood of the everlasting covenant, make you complete in every good work to do His will, working in you what is well pleasing in His sight, through Jesus Christ, to whom be glory forever and ever. Amen" (Heb 13:20, 21).

Thus, all generations and nationalities of those who receive God's free gift of salvation are bought and paid for by the precious blood of the Lamb. Those who receive him as Lord and Savior will ultimately enjoy eternal forgiveness and rest with him.

Scales are mentioned several places in the Bible:

1. "A false balance [*Mozen*] is an abomination to the Lord: but a just weight is his delight" (Prov 11:1).

2. "Honest weights and scales are the Lord's; all the weights in the bag are His work" (Prov 16:11).

The governments of the world are to govern according to the supreme government of God: "Let every soul be subject to the governing authorities. For there is no authority except from God, and the authorities that exist are appointed by God" (Rom 13:1).

The Bible says: "Blessed is the nation whose God is the Lord, the people He has chosen as His own inheritance" (Ps 33:12).

Metaphorically, God "weighs" nations and individuals. Nations who welcome his lordship have a greater degree of peace and success. It behooves us to remember, "unless the Lord builds the house, they labor in vain who build it" (Ps 127:1). More precisely, unless the Lord builds the *family, church, career, company, city, and nation*, they labor in vain who build it.

Belshazzar, coregent of Babylon, is an example of a leader who ignored God and pursued his own agenda. For this reason, "the handwriting on the wall" appeared in approximately 553 BC: "You have been weighed in the balances, and found wanting [deficient]" (Dan 5:26–27).

God employs weighing scales to illustrate the sad state of mankind without God in the world. Psalm 62 pronounces the verdict of our lost world having been weighed in the balances and found deficient. "Surely men of low degree are vanity [הֶבֶל/*hebel*], and men of high degree are a lie: In the balances [*Mozen*] they will go up; they are together lighter than vanity [הֶבֶל/*hebel*]" (Ps 62:9 ASV).

The Hebrew word "vanity" is הֶבֶל/*hebel*. Solomon used this word five times in the opening chapter of the book of Ecclesiastes regarding the condition of man: "'Vanity of vanities,' says the Preacher; 'Vanity of vanities, all is vanity'" (Eccl 1:2).

Adam and Eve named their second son Abel (Hebel). The name "Abel" paints a bleak picture of the deficient condition of the emerging human race. Yet, there was hope for Adam and Eve through the first messianic prophecy spoken by our Creator, Redeemer in Genesis: "And *I will put* [שִׁית/*shith*] enmity between you and the woman, and between your seed and her Seed; He shall crush your head, and you shall bruise His heel" (Gen 3:15).

The Hebrew word for "I will put" is שִׁית/*shith*, which means "to put, apply, appoint." Adam and Eve apparently named their third son Seth after this prophecy. The name שֵׁת/*Sheth* is from *shith*: "And Adam knew his wife again and she bore a son and named him Seth [שֵׁת/*Shêth*] For God has appointed [שִׁית/*shith*] another seed for me instead of Abel, whom Cain killed" (Gen 4:25).

Adam and Eve knew that the way out of the dilemma of death they—and the human race—were facing would be the fulfillment of the promised Messiah through the lineage of Seth. Seth is the appointed off-spring, the seed of promise, which led to the Messiah. Adam and Eve were in essence looking for and hastening unto the first coming of the Son of God.

Interestingly, Flavius Josephus, the first-century Jewish historian, traces the origin of astronomy to Seth, the son of Adam: "They [Seth and his family] also were the inventors of that peculiar sort of wisdom which is concerned with the heavenly bodies, and their order."[2]

"God afforded "these ancients" a longer time of life on account of their virtue, and the good use they made of it in astronomical and geo-metrical discoveries."[3]

Four thousand years after Adam and Eve sinned the Gen 3:15 prophecy was fulfilled by the Messiah. When Yeshua died on the cross, he dealt a death blow to Satan's head. This "death blow" is the "eternal, mortal death blow" to Satan's authority over the redeemed.

Messiah's heel was bruised on the cross. There may have been a piece of wood at the lower part of the cross, under the nailed feet of Yeshua, which he may have used as a means to stabilize himself. Pushing himself upward, he gasped for one last breath, gaining strength enough to push, symbolically, downward on Satan's head. Messiah suffered an excruciat-ing, yet temporary, *physical* death, on our behalf. Today, the only way we can enjoy kingdom authority is to receive *the* king: King Jesus!

Today, six thousand years after Adam and Eve sinned, we're antici-pating *the final blow* to Satan. The Messiah is coming back to *permanently* remove Satan, the author of hate, from the people of God and creation itself.

The nature of God is love. The love of God is weighty. The fruit of the Spirit is weighty and produces worthy character in us. God desires our worthy character and works be placed on his scales of judgment. The only way to secure this "worthy weight" is through a personal relation-ship with the one who carried the weight of the cross on our behalf. We must *know* Yeshua. It is he who crushed the dragon's head—releasing us from Satan's power and authority.

Yeshua redeems us from vanity. When one comes to God in the name and blood of Yeshua, he forgives our sins and removes them from

2. Josephus, *Works of Josephus*, 1.2.68–71.
3. Josephus, *Works of Josephus*, 1.3.106.

the scales of judgment. *The blood of Messiah, and his accredited righteousness, neutralize our weightless works of vanity.*

The book of Ecclesiastes begins by saying, "Vanity of vanities, all is vanity." It concludes by saying: "Let us hear the conclusion of the whole matter: Fear God and keep His commandments, for this is the whole duty of man. For God will bring every work into judgment, including every secret thing, whether good or evil" (Eccl 12:13–14).

Solomon proved that vanity does not work. Solomon, the richest man in the world, was attempting to prove or disprove the value of vanity. He found "vanity" to be futile and frustrating, and going nowhere—investing in temporary things of no "worthy weight." Worthy weight is gained by those who receive God's gracious gift of salvation and live according to his word.

People without Christ, known as "the world," attempt to save their lives through vanity. The Bible teaches that we must lose our life to find life in the weighty love of Messiah. Only in him can we find new life, and appropriate his character: hope, love, peace, and works. Christ's characteristics in us is the eternal weight of glory.

When we appear before the judgment seat of the Messiah, every thought, word, and deed will be "tried by fire." Yeshua said that even giving a cup of water to someone in his name will be rewarded. Rewards are based on what we truly do in Yeshua's name. Yeshua's name denotes authority and honor. "And whatever you do in word or deed, do all in the name of the Lord Jesus, giving thanks to God the Father through Him" (Col 3:17).

The apostle Peter wrote: "As each one has received a gift, minister it to one another, as good stewards of the manifold grace of God. If anyone speaks, let him speak as the oracles of God. If anyone ministers, let him do it as with the ability which God supplies, that in all things God may be glorified through Jesus Christ, to whom belong the glory and the dominion forever and ever. Amen" (1 Pet 4:10, 11). The application of this scripture produces good success!

Words and works done in Yeshua's name are substantive and reap eternal rewards. This "weight" secures a crown before the judgment seat of the Messiah, which we can cast before his feet in everlasting gratitude and worship. Each person's crown is indicative of the constellation Corona which represents the all-encompassing crown of Messiah. Corona is a symbolic representation of dominion lost and dominion found. Each crown is reflective of the crown of the Messiah, the King of kings, and LORD of Lords.

Vain words and works of the world are of no substantive value to God and will be burned. Yeshua said: "But I say to you, that for every idle word men speak, they will give account of it in the day of judgment. For by your words, you will be justified, and by your words you will be condemned" (Matt 12:36, 37). The Greek word *idle* means "inactive, useless, unprofitable, barren."

As children of Adam, our Creator, Redeemer saw great value in us. So much so, that the he was willing to lay down his life for his creation. "For God so loved the world that He gave His only begotten Son, that whoever believes in Him should not perish but have everlasting life" (John 3:16).

Yeshua explains the value of the children of Adam in his parables of the hidden treasure and pearl of great price: "Again, the kingdom of heaven is like treasure hidden in a field, which a man found and hid; and for joy over it he goes and sells all that he has and buys that field" (Matt 13:44). "Again, the kingdom of heaven is like a merchant seeking beautiful pearls, who, when he had found one pearl of great price, went and sold all that he had and bought it" (Matt 13:45, 46).

God's people Israel/the church are, collectively, like a treasure found and purchased by our great Redeemer. Each person chosen by God is like a pearl of great price for whom the Son of God laid down his life. Out of an unclean oyster comes a pearl of great price. Pearls and treasure are valuable when weighed in the balances. Yet, even a feather has weight on the weighing scales. Vanity has none.

Libra is an emblem of justice for all. Mankind is accountable to the supreme court of heaven for their sins. Sin is the violation of God's commandments. "Whosoever commits sin transgresses the law: for sin is the transgression of the law" (1 John 3:4).

God takes ultimate responsibility for the fall of mankind. He came to provide pardon for the children of Adam. Man could not redeem (purchase) himself from the fall. Only God can redeem (purchase) man. In order for the God of the universe to be just and the justifier of those who receive Yeshua, God become a man. In order to fulfill the just penalty for sin, the Lord of creation carried the weight of a Roman cross, was crucified, and shed his perfect blood on our behalf to satisfy the just demands of his perfect law.

Our most liberating revelation is knowing our sins were nailed to the cross with the Messiah. The book of Colossians expounds on the fact that the very *record* of our sins has been blotted out: "And you, being dead in your trespasses and the uncircumcision of your flesh, He has made alive together with Him, having forgiven you all trespasses, having wiped out the handwriting of requirements that was against us, which was contrary to us. And He has taken it out of the way, having nailed it to the cross" (Col 2:13, 14).

The Greek word for "handwriting of requirements" χειρόγραφον, ου, τό/*cheirographon* means "certificate of debt, or legal note, the record of our sins." The "record book of sins" was placed in a courtroom before the Judge's bench, as a witness against the wrongdoer. The "record book" of our sins has already been expunged by our attorney, Messiah Yeshua. This "record book" has been, figuratively, nailed to the cross with Messiah (Col 2:13–14; 1 John 2:1). God has so fully forgiven his people that not only are our sins forgiven, the very record of our sins has been removed. In the Messiah, we have a clean slate, a brand new beginning. This complete *justification*, "*just-as-if*" *I never sinned*, enables us to live and work based on God's verdict of freedom. We are free to go forth and serve him!

Our Savior gives us a new heart. This new "heart" (see Gal 5) enables us to love him and keep his commandments. He teaches us to live a *balanced lifestyle* as displayed in *Libra, the scales*. Not a balance of good and evil as some religions teach, but to live our lives in balance with God's standards. God wants the weight of his righteousness working in us and through us.

Five is the number of grace. Below are five major parts to the weighing scales:

1. the fulcrum: the vertical point, or head post on which the balancing arms rest

2. the vertical beam represents God's straightforward fairness

3. the horizontal beams of equal length represent the arms of the Lord

4. the right weighing pan signifies the Ten Commandments—righteous (known weight)

5. the left weighing pan signifies one's moral position and practice (unknown weight)

The star Zuben El Genubi corresponds to the right weighing pan, meaning "scale south" or "southern scale." I am proposing that the (known) weight placed on the right weighing pan represents the Ten Commandments. The Ten Commandments are God's perfect, eternal, absolute, and unchanging moral standard for mankind.

The star Zuben Al Chemali corresponds to the left weighing pan, meaning "scale north" or "northern scale." The (unknown) weight placed on the left weighing pan identifies one's moral position—whether the individual is *in* Adam or *in* the Messiah. This "unknown weight" represents one's nature, character, and works being weighed in comparison to God's unchanging eternal law.

The objective of this scale is to apply weight to the left weighing pan that equals the weight of the right weighing pan. This object lesson illustrates that God is searching for character and works based on his absolute goodness and universal law. Are we living our lives in balance with the righteous requirements of his love and law? God desires the right and left side of his weighing to be equally balanced. He wants mankind to live according to *his* divine, universal love and law—which is just, holy, and good. Living a balanced life can be realized only in the Messiah. Each day we are given an opportunity to live a balanced life. We can embrace this opportunity by choosing to walk in agreement with our Creator and Redeemer through the application of his word.

Libra's "balanced scales" are a picture of God's investment in his people. He requires our faithfulness and honesty in return. Yeshua spoke the parable of the talents as the "standard," in keeping with the scales: "For the kingdom of heaven is like a man traveling to a far country, who called his own servants and delivered his goods to them. And to one he gave five talents, to another two, and to another one; each according to his own ability; and immediately he went on a journey" (Matt 25:14, 15).

In ancient Greece, a talent was the standard unit of weight used to balance a two-armed weighing scale. An object lesson of the weighing scale can be compared to God placing talents or "weights" in us. Having been purchased by God, through faith in Yeshua, may we live our lives in humble gratitude for the costly investment God has made in us.

The apostle Paul reminds us of God's investment: "But we have this treasure in earthen vessels, that the excellence of the power may be of God and not of us" (2 Cor 4:7). The Greek word for treasure is θησαυρός, οῦ, ό/*thésauros*, meaning "a deposit, wealth (literally and figuratively); treasure—a receptacle for valuables."

The left weighing pan represents the Lord's glory restored in his people. The Hebrew word for glory is כָּבוֹד/kä·vōde from כָּבַד/kä·vad', meaning "to be heavy, weighty (but only figuratively), in a good sense, splendor or copiousness—glorious, glory, honor."

The apostle Paul said: "For our light affliction, which is but for a moment, is working for us a far more exceeding and *eternal weight of glory*, while we do not look at the things which are seen, but at the things which are not seen. For the things which are seen are temporary, but the things which are not seen are eternal" (2 Cor 4:17–18).

The apostle Paul reminds us: "for all have sinned and fall short of the glory of God" (Rom 3:23). In Messiah, we are restored to God's glory. The fruit of the Spirit is an expression of the nature, character, and glory of the Lord in us.

For example, when Moses said to the Lord, "Please let me see Your glory," the Lord said, "I will make all My goodness pass before you, and I will proclaim the name of the Lord before you. I will be gracious to whom I will be gracious, and I will have compassion on whom I will have compassion" (Exod 33:18, 19).

Following this encounter, the Lord said to Moses, "Cut two tablets of stone like the first ones, and I will write on these tablets the words that were on the first tablets which you broke" (Exod 34:1).

The Ten Commandments are a transcript of the Lord's character and glory. When we are restored to the glory of God in Messiah, we are restored to his character and commandments.

As individual members of the body of Messiah, God has invested measures of grace and glory in each of us. The Lord desires to conform us to the image of Messiah, to grow daily in his likeness. As displayed in the constellation Libra, this growth is depicted by moral maturity and a godly lifestyle.

The cross of Christ is the ultimate expression of the vertical and horizontal love of God. The cross exhibits the *Messiah's* vertical love to the Father, and his horizontal love to the world. The cross exhibits *our* vertical love to God and our horizontal love to our neighbor.

When we agree with God's covenant and commandments, we discover a life of equilibrium. As a child, I remember walking on railroad tracks balancing myself by extending my arms, on each side of my body, to achieve equilibrium. Walking requires balance. Everything is based on balance. Seeking first the kingdom of God and his righteousness is a matter of balance. To work six days and rest on the seventh day is a matter of balance. The Bible says: "Who has measured the waters in the hollow of His hand, measured heaven with a span and calculated the dust of the earth in a measure? Weighed the mountains in scales and the hills *in a balance*?" (Isa 40:12).

Sin, the violation of God's commandments, is represented as "unjust" weights. Accordingly, there should be only "just" weights on the weighing pans. There should not be a combination of just and unjust weights on the scales. The scripture says: "Let us lay aside every weight, and the sin which so easily ensnares us, and let us run with endurance the race that is set before us, looking unto Jesus, the author and finisher of our faith, who for the joy that was set before Him endured the cross, despising the shame, and has sat down at the right hand of the throne of God" (Heb 12:1, 2).

As Yeshua's disciples, we must remember our old nature was cruci-fied with the Messiah on the cross (Rom 6; Gal 2:20). When we reckon the old man (the old nature) dead, worthless and weightless, vanity is removed from the (left) weighing pan. Reckoning our old nature as dead, our "weight of vanity" is neutralized; therefore, our new "weighty" na-ture increases, in the Messiah. Furthermore, the Greek word "reckon" is λογίζομαι/*logízomai*, which means "to consider, take into account, cal-culate, number, weigh." The power of the cross neutralizes the weight of vanity in our lives and energizes the weight of God's glory.

It is essential that we understand what sin is and what sin is not ac-cording to the Scriptures. The Bible alone defines sin and righteousness. When we understand the difference, we can choose to activate the law of the Spirit of life in the Messiah and not activate the law of sin and death in our lives. "Whosoever commits sin transgresses the law: for sin is the transgression of the law" (1 John 3:4).

The Hebrew word for law is תּוֹרָה/*torah*, which means "law, direc-tion, instruction" with the word אוֹר/ '*ôr*, "light," as contained in Torah. "My tongue shall speak of Your word, for all Your commandments are righteousness" (Ps 119:172). "The righteousness of God, through faith in Jesus Christ, to all and on all who believe" (Rom 3:22).

To those *in* Adam, the Ten Commandments are called the law of sin and death. This simply means that sin, through the law of cause and effect, brings death. To those *in* the Messiah, the Ten Commandments are the law of the Spirit of life. In other words, the cause and effect of Yeshua's righteousness in us brings life, thus enabling us to keep his command-ments (Rom 6, 7, 8).

God's law remains steadfast for those living in Messiah and those living in Adam. It is the same law but functions with opposite outcomes. For those in Messiah, the outcome is living in the grace of God, which brings liberty in accordance to the perfect law of liberty (read Jas 1:25; 2:12). For those in living in Adam, the outcome is living "under the penalty of the law."

There is a difference between sin and *sins*. Sin is "inherited." We are *born into* sin through Adam's nature. A tree trunk, for example, represents one's nature, the branches bearing the fruit of its nature. The Bible teaches that those in Adam have a fallen (sin) nature which places them under the penalty of the law of sin and death.

Sinners cannot balance the scales of justice. Only God imparts the weight of righteousness and grace to an individual in balancing his scales of justice. His impartation of pardon, mercy, grace, and righteousness is realized only upon one's acceptance of God's gift of salvation. This is not to say a sinner cannot do "good." Sinners can "do good things." Sinners cannot, however, redeem themselves from their fallen nature. Only Yeshua can liberate man from his fallen Adamic nature, thus enabling him to live a balanced, abundant, and victorious life. This is the gospel according to the Bible, as revealed in the night sky, through the constellation Libra.

As believers in the Messiah, it is important to remember that we are created in the Messiah for good works (Eph 2:10). We are not saved *by* works, but God will judge our works at the judgment seat of the Messiah. We work not *to be* saved—but because we *are* saved.

The apostle Paul encourages us to bring worthy weights to the judgment seat of the Messiah:

> But let each one take heed how he builds on it. For no other foundation can anyone lay than that which is laid, which is Jesus Christ. Now if anyone builds on this foundation with gold, silver, precious stones, wood, hay, straw, each one's work will become clear; for the Day will declare it, because it will be revealed by fire; and the fire will test each one's work, of what sort it is. If anyone's work which he has built on it endures, he will receive a reward. If anyone's work is burned, he will suffer loss; but he himself will be saved, yet so as through fire. (1 Cor 3:10–15)

May we work in response to his love and investment characterized by gold, silver, and precious stones—not wood, hay, and straw. Gold, silver, and precious stones endure the fiery judgment—but wood, hay and straw will be burned. Gold, silver, and precious stones symbolize (unknown) worthy weights placed on the left weighing pan. Wood, hay, and straw symbolize (unknown) unworthy weights placed on the left weighing pan. Gold represents works tested by fire and found to be pure as a result of being partakers of God's divine nature. Silver represents redemption and works purified by the refiner's fire, redeeming us from dead works to living works in the Messiah (Mal 3:2, 3). Precious stones may represent the variety of good works in one's life. Wood, hay, and straw represent works done in one's own strength or with wrong motives. These works will be burned in judgment. God graciously allows these works to be burned from our memories.

We cannot take anything with us when we enter the world to come except our Christ-like character and works done in Yeshua's name. One of the greatest treasures we take with us, in the world to come, are the souls we win to the Messiah. May the fruit of our lives bring joy to the heart of our Creator, Redeemer as we're weighed on the scales of judgment!

The message of Libra communicates that Adam's race cannot redeem itself by "works." Only the Messiah's substitutionary death on the cross grants the worthy weight of perfection which pays our sin debt and affords our redemption. Only Yeshua and the Holy Spirit in us can enable us to keep his commandments and do those things which are pleasing in his sight (1 John 3:22).

The Messiah's blood justifies us before the throne of God. Halleluyah for the worthy weight of Yeshua. God the Father has attributed worthy weight to us in the Messiah. As we grow in him, he imparts to us everything we need to live according to his lifestyle. He guides us into all practical truth which brings freedom and equilibrium. Let us learn to live, clothed in the heavenly armor of Messiah. His armor is worth its weight into which no weapon can penetrate.

Remember, when it comes to salvation, Yeshua, figuratively, stands on the "weighing scale" on behalf of his redeemed people (Eph 2:8–9). When it comes to rewards, Yeshua lives his life of righteousness through us as we learn to appropriate his equilibrium (Rom 8:1–4).

Notes on My Sketch of Libra

Conventional imagery depicts Libra as having a base-style scale. The group of three stars located at the center bottom of the constellation align to form a vertical base style scale.

Ancient imagery depicts Libra as a handheld scale. Historical examples of Libra depicted as a handheld-style scale include:

1. The ancient four-thousand-year-old Dendera Egyptian Zodiac.

2. The stunning unearthed mosaic floor of a fourth-century synagogue in Tiberias, Israel.

3. The splendid mosaic zodiac floor of the sixth-century AD Beth Alpha Synagogue.

4. The mosaic floor depicting the zodiac at the fifth century Tzippori Synagogue.

My sketch confirms Libra's design based on the star positions as documented in Stellarium software. Stellarium calculates the positions of the stars and shows a realistic sky in 3D just as it appears with the naked eye, binoculars, or telescope.

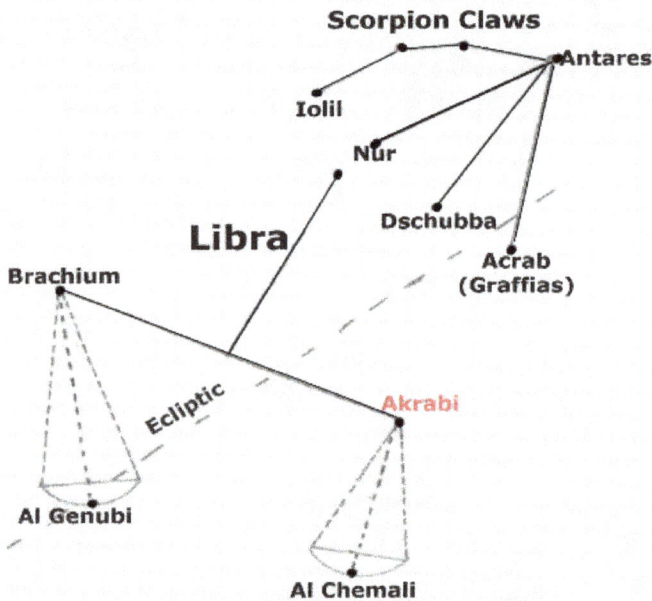

My sketch (design) of Libra is based on tracing the star positions with Stellarium software.

My star song for Libra is titled "The Price of Redemption."

The Price of Redemption

By Bruce J. Patterson

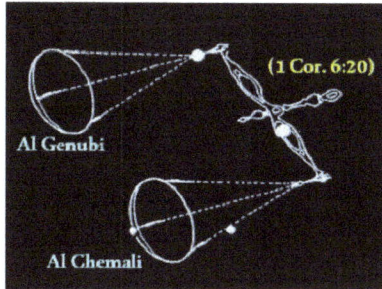

Verse 1

The price deficient was tilted high
So high, so high
Weighed in the scales, found wanting
Wanting, wanting
Who would come to pay the price?
God's own Son by his sacrifice
He paid the penalty for man's sin
Redemption's price outweighed the price deficient
By the price that covers

Chorus

The price of redemption would require death to satisfy
The holy supreme justice of the God of the universe
The price of redemption to ransom man
Propitiation by God's own hand
The price of redemption he paid

Verse 2

Man in his depravity
could not redeem
What was lost nor pay the cost
to be redeemed
So the grace of the Branch would come
To purchase us by his cross
Where God's mercy and justice met

For the price of the conflict He was sent, Yeshua
His blood now covers

Chorus / Bridge
Where do the balances lean in your life?
The weight is just for the redeemed
In God's holy sight
Oh exchange your deficiency for his sufficiency of price—

Chorus

God's Standard of Righteousness
Justice and Mercy

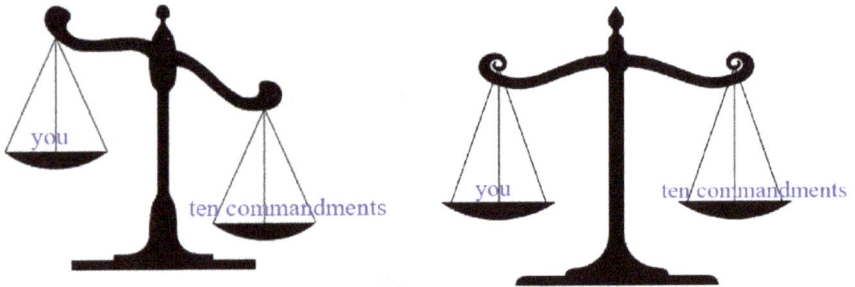

LEFT: The Price Deficient (in Adam) RIGHT: The Price Which Covers, the
Price of Redemption, the Price of the Conflict (in Messiah) (Scale images by
Gospel Shaped Family, info@gospelshapedfamily.com.)

Constellation Name	Translation
Libra (Latin)	balances, scales, a pair of scales
Moznayim (Hebrew)	scales, Hebrew name of Libra constellation
Al-Mizan (Arabic)	the balance
Zyygos (Greek)	the scales
ZI.BA.AN.NA (Sumerian)	the scales, balance
Zibānītu (Akkadian)	balance

Constellation Name	Translation
ma ' sa: tha (Assyrian)	Libra
Mīzān (Persian)	a balance, pair of scales
Lambadia (Persian)	station of propitiation (from *lam*, graciousness, and *badia*, branch) W. E. Bullinger definition corrected: (from *lam*, forgiveness and rest and *badia*, eternal) inferred: station of propitiation

Enclosed are derivatives and similar words in ancient languages which help clarify the core meaning of this constellation/star name.

Ancient Hebrew/ Aramaic	Phonetic Spelling	Summarized Meaning
Hebrew מֹאזְנַיִם	*moznayim*	scales
Hebrew מֹאזֵן	*mô'zên*	(scales) from אָזַן/ *'âzan* (to weigh)
Aramaic נֵים מֵאזֵן	*nasar*	a pair of scales
Aramaic	*zbwn, zibbūn, zibbūnā*	purchase
Hebrew נִשְׁלָם	*nishlam*	completed (John 19:30)

Ancient Gentile Language	Phonetic Spelling	Summarized Meaning
Latin	*Lībra*	scales
Arabic	*mīzān*	scales
Persian	*mīzān*	scales
Assyrian	*zibanitu*	scales
Akkadian	*zibānu*	scales
Sumerian	*ŠÈ-ba-an*	a measure
Persian	*lam*	mercy, forgiveness; rest; tranquility
Persian	*abadī*	eternal; for ever and ever
Persian	*abadīyat*	eternity
Arabic	*abadiyy*	eternal (also: endless, incurable, everlasting, timeless)

Ancient Gentile Language	Phonetic Spelling	Summarized Meaning
Egyptian Arabic	*labaaqa*	grace
Egyptian Arabic	*bae'i*	(the) rest
Egyptian Arabic	*abad*	eternity
Assyrian	*a: ba:d*	forever, eternity, forever and a day
Turkish	*ebediyet*	eternity
Arabic	*ẓufr*	claw
Arabic	*zubanāh*	the claws
Persian	*ẓufr*	claw
Persian	*zubāniyāni*	the two claws (of a scorpion)

Scripture References

Lev 19:36; Job 6:2, 31:6; Ps 62:9; Prov 11:1, 16:11, 20:23; Isa 40:12; Jer 32:10; Amos 8:5; Mic 6:11; Dan 5:27; 1 John 2:1

The Stars

Zuben El Genubi

THE STAR NAME ZUBEN El Genubi (pronounced, zoo-BEN-el-je-NEW-bee) means "scale south" or "southern scale." Zuben El Genubi is a combination of three words: (1) *Zuben*, meaning "scales" in Akkadian. (2) *El* is the Arabic definite article "the." *Genubi* means "south" in Arabic and Persian. Zuben El Genubi, therefore, means, "the southern scale."

Zuben El Genubi corresponds to the right pan of the weighing scales.

The book of Deuteronomy states weights are made of stone: "You shall not have in your bag differing weights [אֶבֶן/*eben*, 'a stone'], a heavy and a light. You shall not have in your house differing measures [אֵיפָה/*ephah*, 'a measure of grain'], a large and a small. You shall have a perfect [שָׁלֵם/*shalem*, 'complete, safe, at peace, perfect' and 'just,' צֶדֶק/*tsedeq*] righteous weight, a perfect and just measure, that your days may be lengthened in the land which the Lord your God is giving you. For all who do such things, all who behave as the unrighteous, are an abomination to the Lord your God" (Deut 25:13–16).

The *eben* stone was placed on one pan of the scales to establish the standard weight. The measure of the item being purchased was added to the other pan until the pans were as closely balanced as possible.

The (known) weight placed on the right weighing pan represents God's eternal unchanging law and standard for man, the Ten Commandments, which are just, holy, and good (Rom 7:12).

The (unknown) weight, placed on the left weighing pan, represents one's moral position in Adam or the Messiah and one's works being weighed compared to God's unchanging eternal law.

For those in the Messiah, the right weighing pan going down (south) symbolizes that the weight of God's law is great. But there is a lack of responsibility to carry out the Lord's righteousness requirements in one's life, thus illustrated by the left weighing pan rising (north). This shows imbalance, or being found lacking. God does not require something from us for which he hasn't equipped us to do or to become. God would be unjust to require of us something we could not become and carry out. The apostle Paul said: "I can do all things through Christ who strengthens me" (Phil 4:13).

For those in Adam, the right weighing pan going down (south) symbolizes the weight of God's law is great. In this case there is no content of worthy substance on the left weighing pan, so the left side rises and the right side goes down. This demonstrates that there is a lack of ability to carry out the Lord's righteous requirements in one's life. The only way one is able to do what is right in God's sight is through God's grace, favor and faith in the Messiah.

Man—separated from God—does not have the ability to fulfill God's commandments: "Because the carnal mind is enmity against God; for it is not subject to the law of God, nor indeed can be. So then, those who are in the flesh cannot please God" (Rom 8:7, 8).

God desires that the right and left side of the scales be equalized, counterbalanced. He wants us to live in accordance with his divine universal law.

Genubi is a derivative of the Hebrew גָּנַב/*ganab*, meaning "to steal, to take away by theft secretly." Yeshua said: "The thief does not come except to steal, and to kill, and to destroy. I have come that they may have life, and that they may have it more abundantly" (John 10:10).

The claws of Scorpio, the Scorpion, are depicted "grasping and tilting" the scales of Libra. Grasping and tilting indicates an attempt by the enemy to steal (*ganab*) truth and balance from mankind. The enemy's objective is to pervert justice and truth through agents of Satan.

Those in Messiah realize that Yeshua plucked us from the claws of the enemy. Yeshua died as a thief in order to snatch us away from Satan, "the thief," the enemy of our souls. The lover of our souls delivers us from the enemy of our souls.

When we invite Yeshua into our lives, he lives his life of righteousness through us that we might learn to appropriate his "balance" or equilibrium.

Note: According to scholarly consensus, the Messiah was crucified on Passover (Nisan 14), Wednesday, April 25, AD 31. He rose from the dead in newness of life, Sunday morning, April 29. Using astronomy software, I discovered amazing astronomical signs in the heavens over Jerusalem when reversing the universe to Wednesday, April 25 and Sunday morning, April 29, AD 31.

Epic Discovery

A full lunar eclipse/blood moon occurred (in Libra the Scales and claws of the Scorpion) on Passover, April 16, 1448 BC at midnight. This Passover blood moon may have occurred during the exodus or one year after the exodus, when Israel was at Mount Sinai in Arabia.

On April 25, AD 31, the night following the crucifixion of the Messiah, the Moon virtually rose (at 7:00 p.m.) in the same position in Libra as it did on Passover, April 16, 1448 BC. Two and a half hours later a partial lunar eclipse of the Moon appeared over Jerusalem. This lunar eclipse (blood moon) peaked at 10:31 p.m. The duration of the eclipse lasted 115 minutes: 9:35 to 11:30 p.m. (1 hour and 55 minutes). This partial blood moon could be symbolic of the first-century redeemed remnant of Israel. About a third of the Moon was totally immersed in Earth's shadow, casting a reddish hue. Atmospheric dust may have intensified the redness of the Moon. A blood moon occurs when the Sun, Earth, and Moon align and the Moon passes into Earth's shadow, casting a reddish glow on the Moon.

The prophet Joel prophesied regarding blood moons and solar eclipses; hence, the apostle Peter quoted the following verse on Shavuot/Pentecost fifty days after resurrection of Yeshua: "I will show wonders in heaven above and signs in the earth beneath: Blood and fire and vapor of smoke. The sun shall be turned into darkness, and the moon into blood, before the coming of the great and awesome day of the Lord" (Acts 2:19–20; Joel 2:28–32).

Joel's prophecy, quoted by the apostle Peter on Shavuot/Pentecost, indicates a blood moon was seen within the context of the Passover crucifixion. This blood moon lunar eclipse occurred April 25, AD 31. Peter's testimony helps identify AD 31 as the year of the Messiah's crucifixion. Note: Passover lunar eclipses (blood moons) of AD 32 and AD 33 were not visible from Jerusalem. The only visible Passover lunar eclipse from

AD 27 to AD 34 is the AD 31 partial blood moon, as seen from Jerusalem (see "Chart of Passover Crucifixion Dates," Appendix 1).

In biblical astronomy, the Moon represents the church, "the spiritual Israel of God." The Moon has no light of its own but reflects the light of the Sun. Likewise, we are to reflect the light of Yeshua (the Sun of Righteousness) in a dark world. Blood moons may represent the blood of the martyrs. Notably, Passover blood moons represent the atoning blood of the Messiah covering and redeeming his people. A blood moon, in the Scales and claws of the Scorpion, represents the Creator, Redeemer snatching his people from the claws of the enemy.

A full lunar eclipse/blood moon occurred in Libra, the Scales, and claws of the Scorpion during what is known as the *Second Passover, May 16, 2022.*

The God of the universe fashioned the constellation Libra to convey his message of justice and mercy. Blood moon lunar eclipses have appeared in Libra, the Scales, and claws of Scorpio, during notable times in history.

God is the author of freedom. As God's people are led by the Holy Spirit and appropriate his truth, greater freedom is experienced in every area of life. God's truth brings security and stability to our lives, enabling us "to prosper and be in health even as our soul prospers" (3 John 1:2).

In summary, the star Zuben El Genubi is a depiction of the right weighing pan and the standard weight of the Ten Commandments it (figuratively) upholds. The Ten Commandments are the summation of God's word and eternal standard for mankind.

As one come to terms with the balance of truth, it is of utmost importance that one grasps the truth of the deity of the Messiah as one of the foremost doctrines of the Bible. We must understand that when God said, *"Let there be light,"* it is he, the agent of creation—the Word made flesh—who *is* the light (Gen 1:3).

It was Yeshua, Jesus, the Word of God, who became flesh, who wrote the Ten Commandments with the finger of God at Mount Sinai (Exod 31:18). It was the Word who became a Jewish man according to John 1:14. It was the Word who stooped down and wrote on the ground with the finger of God (John 8:6, 8). It is the Word who writes the commandments of God in our hearts and minds today (Hebrew 10:16). It is he, whose "robe is dipped in blood, and His name is called 'the Word of God'" (Rev 19:13). Yeshua is the Torah (the law) made flesh, who says to us: "If you love Me, keep My commandments" (John 14:15).

Conventional Star Name/Meaning	Name/Meaning Confirmed or Corrected
Zuben El Genubi—the price which is deficient	corrected: scales tilted southward

Enclosed are derivatives and similar words in ancient languages which help clarify the core meaning of this constellation/star name.

Ancient Hebrew/ Aramaic	Phonetic Spelling	Summarized Meanings
Hebrew מֹאזְנַיִם	mozen	scales
Aramaic	mo'zen	scales
Aramaic	kprh	atonement
Aramaic	zbwn, zibbūn, zibbūnā	purchase
Hebrew גָּנַב	ganab	to steal
Hebrew גֻּנַּבְתִּי	gə·nub·tî	[whether] stolen
Aramaic ܓܢܒ	G'aNeB	steal
Aramaic	gnūbū	theft
Hebrew גָּנַב	ganab	carry away
Aramaic	G'aNeB	steal

Ancient Gentile Language	Phonetic Spelling	Summarized Meanings
Akkadian	zibā nu	scales
Arabic	mīzān	scales
Arabic	janūb	south
Persian	janūban	south
Turkish	güneyde	south
Egyptian Arabic	ganoob	south
Assyrian	gi: ' na: va	steal
Hindi	le jaana	to remove
Sumerian	gána-íl	to carry
Arabic	jānib	side, waist

Ancient Gentile Language	Phonetic Spelling	Summarized Meanings
Egyptian Arabic	*ganb*	side, nearby
Sanskrit	*gaNDa*	side

Scripture References

Lev 19:36; Job 6:2, 31:6; Ps 62:9; Prov 11:1, 16:11, 20:23; Isa 40:12; Jer 32:10; Gen 31:39; Amos 8:5; Mic 6:11; Dan 5:27; Lev 17:11; Rom 3:26

Zuben Al Chemali

The star name Zuben Al Chemali (pronounced zoo-BEN-al-sha-MAH-lee) means "scale north" or "northern scale." Zuben Al Chemali is a combination of three words: (1) *Zuben*, meaning "scales" in Akkadian. (2) *Al* is the Arabic definite article for the word "the." *Chemali* means "north" in Arabic, Persian, and Turkish. Therefore, Zuben Al Chemali means "the northern scale."

Zuben Al Chemali corresponds to the left pan of the weighing scales.

"The (unknown) weight" placed on the left weighing pan represents whether one's moral position is *in* Adam or *in* the Messiah. The word *in* is biblically significant. The word *in* denotes identity and relationship. As descendants of Adam, every human being is born *in* Adam. This is why we must be born again, literally, from above. When we're born from above, we become new creations *in* the Messiah and our new identity and relationship is *in* him.

For those in the Messiah, the left weighing pan represents the imputed righteousness of the Messiah and the beautiful qualities of a newly created spirit and nature. For those in Adam, the left weighing pan represents the inherited unrighteous seed of the serpent and "weightless" qualities of vanity due to a fallen, marred image (1 John 2:16).

We must understand that there are *two* corporate heads on Earth: Adam, the corporate head of the human race, and the Messiah, the corporate head of the new creation race. All who are *in* Adam are partakers of the seed of the serpent. All who are *in* the Messiah are partakers of the Seed of the Woman. These two seeds are opposed, by their very nature, to one another (Gen 3:15; Eph 2:15,16). And each "seed" bears fruit after its own kind.

God does not require of anyone something which one cannot do or become. The only way one is able to do what is right in God's sight is through God's grace, favor and faith in the Messiah.

"For as the heavens are high above the earth, so great is His mercy toward those who fear Him; as far as the east is from the west, so far has He removed our transgressions from us. As a father pities his children, so the Lord pities those who fear Him. For He knows our frame; He remembers that we are dust" (Ps 103:11–14).

Hear what God says regarding the condition of those in Adam: "Surely men of low degree are vanity, and men of high degree are a lie: In the balances [*Mozen*] they will go up; they are together lighter than vanity" (Ps 62:9 ASV).

The Hebrew word "together" is יַחַד/*yachad*, meaning "united, unity." The whole of humanity united is insufficient to bring goodness to the scales.

The term *lighter than breath* means deficient or lacking. In this case, the left pan is lacking good nature and works. The left pan is lacking moral integrity and content of character, while the right pan is weighty with the commandments of God. The results of this evaluation is found lacking. The picture of the left weighing pan tilting northward is a picture of imbalance—a simple matter of cause and effect. There is too much weight on one side and not enough on the other.

The lesson of the scales illustrates the process of redemption. The Bible teaches that God has purchased his people with the precious blood of the Messiah. "For you were bought with a price; therefore glorify God in your body and in your spirit, which are God's" (1 Cor 6:20). The left weighing pan exemplifies both redemption and position for those in the Messiah. The Hebrew name for this star is Kaphar, which means "to cover" or "the atonement." The left weighing pan represents the blood of the Lamb being applied to those who receive God's gracious gift of pardon.

The left weighing pan also represents our corresponding motives and actions being weighed and balanced according to the right pan. This means we are to live our lives in balance with the Lord's will and commandments. The Lord writes his commandments in our hearts and minds so we may learn to love him who first loved us. He gives us grace to love our neighbor as ourselves. Therefore, it is unwise to compare ourselves with others. Rather, we are to compare ourselves to Yeshua, the head (fulcrum) of the church. Yeshua is the personification of God's eternal standard of truth and mercy for humanity.

> For we dare not compare ourselves with those who commend themselves. But they, measuring themselves by themselves, and comparing themselves among themselves, are not wise. (2 Cor 10:12)

Chemali is similar to the Hebrew word שִׂמְלָה/simlâh, or "clothes," specifically, "mantle or covering." Chemali is also phonically similar to the Hebrew word מְעִיל/meʿîyl, meaning "a robe, or mantle" from מָעַל/mâʿal (in the sense of covering for acts of transgression). This is significant, because only the Lord can clothe us in his pure garments of righteousness. The apostle Paul says: "For you are all sons of God through faith in Christ Jesus. For as many of you as were baptized into Christ have put on Christ" (Gal 3:26, 27.)

The star Chemali is also affiliated with the Hebrew word semâ'lîy or "left hand." Chemali corresponds to the left measuring pan of the scales. God wants the right side and left side counterbalanced. He wants us to live in accordance with his divine universal law.

As stated, this star is also known by the Hebrew word kaphar, which means "to cover, atonement." When we receive the Messiah, we are figuratively covered in the robe/meʿîyl of righteousness provided by God.

Yeshua figuratively stands on the weighing pan or scale on our behalf and lives his life of righteousness through us. Through him, we can learn, practice, and appropriate equilibrium.

The book of Revelation depicts the works of the glorious church for whom Yeshua is returning. Our living works count for time and eternity. These works correspond to the (unknown) weight placed on the left weighing pan of the scales—and the rewards it brings. This is a picture of being clothed in a robe of righteousness, the beautiful garment of Messiah. "And to her it was granted to be arrayed in fine linen, clean and bright, for the fine linen is the righteous acts of the saints" (Rev 19:8).

Conventional Star Name/Meaning	Name/Meaning Confirmed or Corrected
Zuben Al Chemali—the price which covers	corrected: scales tilted northward infers "the price which covers"

Enclosed are derivatives and similar words in ancient languages which help clarify the core meaning of this constellation/star name.

Ancient Hebrew/ Aramaic	Phonetic Spelling	Summarized Meaning
Hebrew מֹאזְנַיִם	*mozen*	scales
Aramaic	*mo'zen*	scales
Aramaic	*zbwn, zibbūn, zibbūnā*	purchase
Hebrew שִׂמְלָה	*simlâh*	a dress, a mantle, wrapper, covering garment, garments, clothes, raiment, a cloth perhaps by permutation for the fem. of סֶמֶל/*çemel*, *seh›-mel* (through the idea of a cover assuming the shape of the object beneath); image, statue, idol Gen 3:21
Hebrew מְעִיל	*me῾îyl*	a robe, mantle, from מָעַל/ *mâ'al* (in the sense of covering) used only figuratively, to act covertly, i.e., treacherously: to transgress (commit, to do a) trespass(-ing)
Aramaic	*semmāl, semmālā; smāl, smālā*	left hand
Hebrew שְׂמָאלִי	*se mâ'liy*	left hand
Aramaic ܣܡܠܐ	*SeMaLaA*	left hand

Ancient Gentile Language	Phonetic Spelling	Summarized Meaning
Akkadian	*zibānu*	scales
Arabic	*mīzān*	scales
Arabic	*samāl*	north

Ancient Gentile Language	Phonetic Spelling	Summarized Meaning
Persian	*shamāl*	north
Sanskrit	*saumya*	north
Turkish	*simal*	north
Egyptian Arabic	*shamael*	north
Arabic	*šamala*	to cover
Arabic	*s̲himāl*	left
Arabic	*als̲h̃ma̲l*	left, northernness
Persian	*shamāl*	left-hand
Assyrian	*sim'ma:la*	left hand
Sanskrit	*saumya*	left hand
Akkadian	*ša-šumēli*	left-handed

Scripture References

Lev 19:36; Job 6:2, 31:6; Ps 62:9; Prov 11:1, 16:11, 20:23; Isa 40:12; Jer 32:10; Amos 8:5; Mic 6:11; Dan 5:27; Isa 61:10; Rev 19:8

Zuben Akrabi

The star name Zuben Akrabi (pronounced zoo-BEN-AK-ra-bi) means "scales of the conflict or war." Zuben Akrabi is a combination of two words, *Zuben*, meaning "scales" in Akkadian, and *Akrabi*, meaning "conflict, war."

The Hebrew word for Akrabi, קְרָב/*qᵉrâb*, means "conflict, war." Akrabi also means "scorpion," עַקְרָב/*aqrâb*, in Hebrew. This word is appropriate, in that the scorpion (Scorpio) represents the devil.

In the heavens, we see the constellation Ophiuchus, the "Serpent Restrainer," engaged in warfare with the scorpion. Zuben Akrabi reminds us of the price Messiah paid to engage the enemy on his own turf in order to purchase our redemption.

The claws of the scorpion, as depicted in Akrabi, are tilting the scales of Libra. This indicates an attempt by the enemy to steal (*ganab*) truth and balance from mankind. Those in Messiah know that Yeshua plucked us from the claws of the enemy. Yeshua, in essence, died as a thief in order to "steal" us from the enemy's domain. In essence, the lover of our soul delivered us from the enemy of our soul.

The usage of the word *q^erâb* is found in Zechariah: "Then shall the LORD go forth, and fight against those nations, as when He fought in the day of battle [*q^erâb*]" (Zech 14:3). The Messiah will finish the final battle when he returns to destroy those who come against Jerusalem, Israel. And the time is near.

Remember, Christ won the war so we can win the battles we face.

> The thief does not come except to steal, and to kill, and to destroy. I have come that they may have life, and that they may have it more abundantly. (John 10:10)

Conventional Star Name/Meaning	Name/Meaning Confirmed or Corrected
Zuben Akrabi—the price of the conflict	confirmed: scales of war, scorpion inferring: the price of the conflict

Enclosed are derivatives and similar words in ancient languages which help clarify the core meaning of this constellation/star name.

Ancient Hebrew/ Aramaic	Phonetic Spelling	Summarized Meaning
Hebrew קְרָב	*q^erâb*	war
Hebrew עֲקְרָב	*aqrâb*	scorpion
Aramaic ܩܪܒܐ	*QaRB'aA*	war, battle, fighting
Aramaic ܥܩܪܒܐ	*eQaRB,aA*	scorpion
Hebrew נָקוּב	*náqẁb*	perforated, pierced, punctured
Aramaic	*zbwn, zibbūn, zibbūnā*	purchase

Ancient Gentile Language	Phonetic Spelling	Summarized Meaning
Akkadian	*zibānu*	scales
Akkadian	*aqrabu*	scorpion
Akkadian	*araqu*	war

Ancient Gentile Language	Phonetic Spelling	Summarized Meaning
Assyrian	qra: ba	war
Assyrian	eqarba	scorpion
Arabic	taḏāraba	war
Persian	aqrab	scorpion
Turkish	akrep	scorpion
Egyptian Arabic	aa'rab	scorpion
Akkadian	nakadu	sting

Scripture References

Lev 19:36; Job 6:2; Job 31:6; Pss 55:18, 62:9; Prov 11:1, 16:11, 20:23; Isa 40:12; Jer 32:10; Amos 8:5; Mic 6:11; Dan 5:27

Brachium

The star name *Brachium* is Latin for "arm." The Greek equivalency of the Latin word brachium, *brachiōn*, means "arm." The Hebrew equivalency for the Greek word *brachiōn* is זְרוֹעַ/*zᵉrôwa*, meaning "arm." The Greek word *brachiōn* is used three times in the New Testament Bible. Its Hebrew equivalency זְרוֹעַ/*zᵉrôwa* is used ninety-one times in the Hebrew Bible.

The star Brachium, in Libra, may symbolize the arm of the Lord. The horizontal arms of a scale must be in perfect balance in order to uphold the right and left weighing pans, equally. In like manner, the Lord's arms are perfectly balanced—and strong enough—to uphold you and me. The Scripture says, "For the arms of the wicked shall be broken, but the LORD upholds the righteous" (Ps 37:17).

"The eternal God is your refuge, and underneath are the everlasting arms [זְרוֹעַ/*zᵉ ᵉrôwa*]" (Deut 33:27).

"The LORD is just in all his ways and kind in all His deeds" (Ps 145:17).

We can depend on the Lord to teach us to rest fully in his loving arms of grace and mercy. Grace is the unmerited favor of God. Grace is the power to do what is right. When we are doers of the word and not hearers only, we discover amazing balance and equilibrium in our lives. For in him we live, move, and have our being (Acts 17:28).

Classic examples of the Greek and the Hebrew word *arm* are as follows: "Who has believed our report? And to whom has the arm [Heb. *z^erôwa*] of the Lord been revealed?" (Isa 53:1). The Greek Septuagint uses the word βραχίων/*brachíōn* for arm.

"That the word of Isaiah the prophet might be fulfilled, which he spoke: 'Lord, who has believed our report? And to whom has the arm [Gk. *brachíōn*] of the Lord been revealed?'" (John 12:38).

In the Hebrew Bible, "arm" (*z^erôwa*) relates to the salvation of the world: "And he saw that there was no man, and wondered why there was no intercessor; therefore, his own arm [*z^erôwa*] brought salvation for him; and his own righteousness, it sustained him" (Isa 59:16).

The most significant teaching regarding Libra, the Scales, is that it clearly represents our redemption, through Christ. The weighing scales are structured in the form of a man with a vertical profile including his head, horizontal arms, and hands. This is a picture of the Messiah, the Seed of the Woman. In the scales, we see the value of a soul. We see Yeshua, with outstretched arms, paying our sin debt and redeeming us as his very own.

Have you believed the gospel report? Have you looked, with the eyes of faith, to see the arm of the Lord reaching out for you? Thank the Lord, he is the author of truth, justice, mercy, salvation, and strength. His arm is being revealed to the nations today.

The book of Exodus gives a pictorial description of this reality: "I will redeem you with an outstretched arm [זְרוֹעַ/*zeroa*] and with great judgments. I will take you as My people, and I will be your God" (Exod 6:6, 7).

No Conventional Star Name/Meaning	Name/Meaning Included
Brachium—(not included in E. W. Bullinger's book)	included: *Brachium*—arm

Enclosed are derivatives and similar words in ancient languages which help clarify the core meaning of this constellation/star name.

Ancient Hebrew/ Aramaic	Phonetic Spelling	Summarized Meaning
Hebrew זְרוֹעַ	z^{e}rôwa	the arm (outstretched) figuratively, force; arm, help, mighty, power, shoulder, strength.

Ancient Gentile Language	Phonetic Spelling	Summarized Meaning
Latin	brachium	arm, the forearm, lower arm
Greek βραχίων	brachíōn	the arm, i.e., (figuratively) strength: arm
Greek βραχίονας	brachiónas	arm, branch, upper arm
Egyptian Arabic	ziraeA	arm

Scripture References

Ps 37:17; Deut 33:27; Exod 15:16, 6:6; Deut 5:15, 26:8; Isa 53:1, 59:16; John 12:38; Acts 13:17; Luke 1:51; Ps 98:1

6

The Constellation: Crux

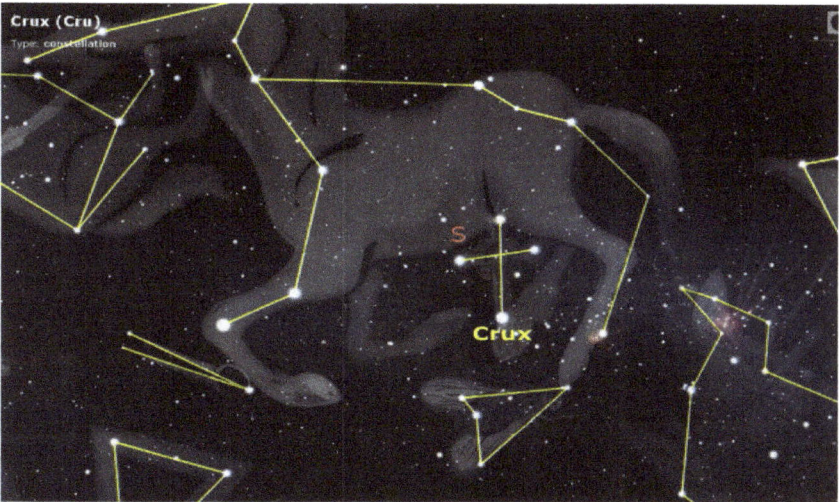

CRUX IS LATIN FOR "cross." Known as "the Southern Cross," Crux is the smallest constellation in the sky. Although its history is ambiguous, Crux is one of the most recognizable constellations in the southern hemisphere. It plays a leading role in the chronicles of Centaurus, the Conquering Horseman, in God's theater in the sky.

Early sightings of the Southern Cross were documented in Greece circa 1000 BC. Greek "star-gazers" associated the celestial cross with the crucifixion of Christ—considering Crux as part of the *Centaurus* constellation.

Today, the International Standard Dating System recognizes "years" according to the traditional "AD" and "BC" system. AD stands for *anno domini*, Latin for "in the year of the Lord," specifically to the birth of Jesus Christ; BC for "before Christ."

The cross of Yeshua is the focal point of history. In the Greek language, Yeshua is the Alpha and Omega. In the Hebrew alphabet, Yeshua is the Aleph and the Tav; the Beginning and the End; the First and the Last (Rev 22:13).

Yeshua was crucified on a Roman cross for the sins of Adam's race. Yeshua is God's greatest expression of mercy and justice. God's mercy and justice (Yeshua) met on behalf of the human race, at the cross, as indicated through the last letter of the Hebrew alphabet, the "*tav*." The early Hebrew *tav* ✝ was shaped like a cross and is defined as a "mark, signature sign, boundary, finished." Similarly, the Egyptian hieroglyph ✝ is a picture of two crossed sticks having the meaning, "mark," "sign" and "signature."

John 19:30 tells us: After Jesus tasted the sour wine (extended to him in a sponge on a branch of hyssop), he cried, "*nishlam/*נִשְׁלָם!" "It is done, completed," in Hebrew.

The Greek translation for "finished," τελέω/*teléō*, is defined as: to end, complete, execute, conclude, discharge (a debt). "When Jesus had received the sour wine, He said, 'It is finished [τελέω/*teléó*]!' And bowing His head, He gave up His spirit" (John 19:30).

The Hebrew name for this constellation is Adom, from אָדֹם *'âdôm*, aw-dome'; meaning "red." This name reminds us of "the blood of his cross" (Col 1:20). Yeshua, the chief Shepherd, purchased his people (flock) with his own precious blood (1 Pet 1:19; Heb 9:22; Acts 20:28).

In his book *The Gospel in the Stars*, 1885, Joseph A. Seiss states, "Formerly, the constellation Crux was visible in our latitudes; but in the gradual shifting of the heavens, it has long since sunk away to the southward. It was last seen in the horizon of Jerusalem about the time Christ was crucified."[1]

Turning the universe back to 9:00 p.m., April 25, AD 31, we observe Crux, the Southern Cross, descending in the southern sky (as seen from Jerusalem) following Yeshua's crucifixion. The southern constellations

---

1. Seiss, *Gospel in the Stars*, 37.

Crux, Centaurus, and Lupus reverberate the universal reality that Christ laid down his life for us.

Crux, truly signifies God's Signature in the Stars.

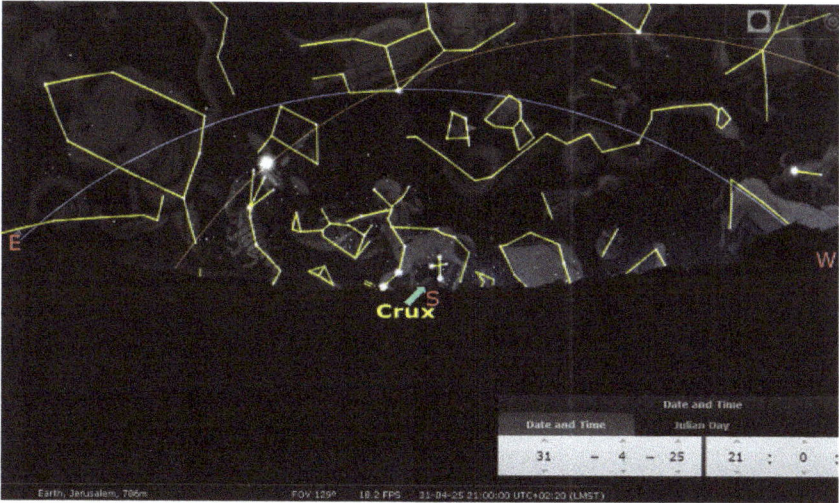

Stellarium close-up image: This astronomical configuration occurred April 25, AD 31, following the Passover crucifixion of Messiah. From Jerusalem, at 9:00 p.m. we see Crux, the Southern Cross, descending in the southern horizon.

My star song for Crux is titled "I Will Glory in the Cross."

I Will Glory in the Cross

(Gal 6:14; Ps 116:12–13)
By Bruce J. Patterson

Crux—Southern Cross. Photo by Chris Picking,
http://starrynightskies.com

Chorus

I will glory in the cross
The token of death to eternal life
I will glory in the cross
Where the Prince of Glory died

Verse 1

God forbid that I should glory
But in the cross of Christ
By whom the world is crucified to me
And I to the life of the world

Chorus

Verse 2

He was pierced for our transgressions
He was bruised for our iniquities
The chastisement that gives us peace
Was upon Jesus

Chorus

Verse 3

What shall I render to the Lord
For all his benefits to me
I will lift the cup of salvation
I will call on the name of the Lord

Verse 4

In the summer sky see the Southern Cross
Up above the world so high
Shinning forth the light of the Lord
And the love of Adonai

Chorus

Constellation Name	Translation
Crux (Latin)	cross
Tav (Hebrew)	mark, sign, a boundary and finished
nán shí zì zuò (Chinese)	the southern cross-shaped constellation

Enclosed are derivatives and similar words in ancient languages which help clarify the core meaning of this constellation/star name.

Ancient Hebrew/ Aramaic	Phonetic Spelling	Summarized Meaning
Hebrew ת	tav	mark, sign, a boundary and finished
Hebrew אָדֹם	âdôm	red, ruddy

Ancient Gentile Language	Phonetic Spelling	Summarized Meaning
Latin	crux, crucis	cross

Scripture References

Matt 27; Mark 15; Luke 23; John 19; Gal 6:14

7

The Constellation: Lupus

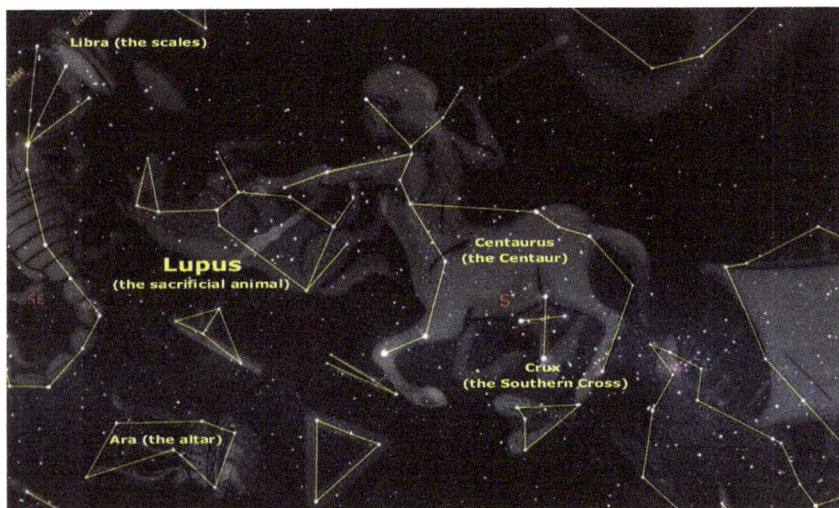

LUPUS IS POSITIONED IN the southern celestial hemisphere of the sky and may be considered a constellation in the *chambers of the south*: "He commands the sun and it does not rise; He seals off the stars; He alone spreads out the heavens, and treads on the waves of the sea; He made the Bear, Orion, and the Pleiades, and the chambers of the south; He does great things past finding out, yes, wonders without number" (Job 9:7–10).

The word *lupus* is the Latin word for "wolf." The constellation, however, is a depiction of a sacrificial animal. According to historians, this constellation was not depicted as a wolf—an unclean, inedible animal—until the sixteenth century. Therefore, "wolf" does not fit the biblical narrative.

Fifteenth century Timurid astronomer, Ulugh Beigh says it was anciently called a sheep or lamb. The Arabs use a word in connection with it which means to be slain, destroyed; hence the slain one, the victim. It plainly expresses slaying, sacrifice by death; Christ is "the Lamb slain from the foundation of the world" (Rev 13:8).[1]

Roman astronomer, musician, and mathematician Ptolemy, and the Greeks, define this constellation as θηρίον/*therion* "a wild animal." They surmise that Centaurus—a constellation appearing as both man and horse—symbolically took this "wild animal" to the altar (the constellation Ara, in Latin). This interpretation reflects its biblical narrative.

In the Greek New Testament, the word θηρίον/*therion* is used in the book of Hebrews in reference to the Jewish people at Mount Sinai in Arabia: "And if so much as a beast [θηρίο/*therion*] touches the mountain, it shall be stoned or shot with an arrow" (Heb 12:20). The Hebrew equivalency of θηρίον/*therion* is בְּהֵמָה/*bᵉhêmâh*, which means "beast, cattle, animal, livestock (of domestic animals), wild beasts." The Hebrew word בְּהֵמָה/*bᵉhêmâh* is used in its original context (Exod 19) as follows:

> You shall set bounds for the people all around, saying, "Take heed to yourselves that you do not go up to the mountain or touch its base. Whoever touches the mountain shall surely be put to death. Not a hand shall touch him, but he shall surely be stoned or shot with an arrow; whether man or beast [מֵהֵב/ *bᵉhêmâh*], he shall not live." When the trumpet sounds long, they shall come near the mountain" (Exod 19:13).

God told the Hebrews not to touch the mountain nor allow the animals to touch the mountain. The animals kept by the Hebrews at Mount Sinai were clean sacrificial animals. For this reason, Moses built an altar at the base of the mountain.

The word בְּהֵמָה/*bᵉhêmâh* includes clean animals in the book of Deuteronomy: "You shall not eat any detestable thing. These are the animals [בְּהֵמָה/*bᵉhêmâh*] which you may eat: the ox, the sheep, the goat, the deer, the gazelle, the roe deer, the wild goat, the mountain goat, the antelope, and the mountain sheep. And you may eat every animal [בְּהֵמָה/*bᵉhêmâh*] with cloven hooves, having the hoof split into two parts, and that chews the cud, among the animals [בְּהֵמָה/*bᵉhêmâh*]" (Deut 14:3–6).

---

1. Seiss, *Gospel in the Stars*, 40.

Of the kosher animals, the *bᵉhêmâh* is limited to three families of animals which were designated by the Torah as animals which could be offered as sacrifices in the Holy Temple during the old covenant dispensation. The Latin terms for these three families of animals are the bovine, the ovine, and the caprine. These Latin-based classifications include cows/bulls, rams/sheep, and goats. These three types of animals were offered as animal sacrifices. The symbolic imagery of the wild ram best fits the narrative within the context of these constellations.

Medieval Christians associated this constellation not with a wolf, but with a ram. They also associated the constellation called "Lupus" with the biblical narrative of Abraham sacrificing his son, Isaac, on Mount Moriah. In Genesis chapter 22, God told Abraham to offer his son Isaac as a burnt offering on Mount Moriah. In obedience, Abraham raised his knife to slay his son. The angel of the Lord, however, stayed his hand. You may recall, Mount Moriah is where the Lord provided a ram caught in the thicket. Abraham slew the ram and offered it as a burnt sacrifice in place of his son, Isaac.

The constellation Centaur portrays Abraham placing Isaac (as the ram) on the altar as a living sacrifice. Ultimately, this "story in the stars" serves as a pictogram of the supreme sacrifice fulfilled by Yeshua, the Lamb of God, who takes away the sin of the world.

The southern constellations Crux, Centaurus, and Lupus reverberate the universal reality that Christ laid down his life for us, and that sin must be judged.

Did you know the cross of Christ is reflected in the adjacent constellation, Crux, the Southern Cross? This sacrificial pictogram is located just below Libra, the Scales, depicting Yeshua balancing "the scales" paying the price of redemption.

The chorus in my star song for the constellation Libra reflects Yeshua's sacrifice:

The Price of Redemption

The price of redemption would require death to satisfy
The holy supreme justice of the God of the universe
The price of redemption to ransom man
Propitiation by God's own hand
The price of redemption he paid

This constellation reveals the Messiah coming as the Lamb (Ram) of God who voluntarily laid down his life for you and me: "Therefore, My Father loves Me, because I lay down My life that I may take it again. No one takes it from Me, but I lay it down of Myself. I have power to lay it down, and I have power to take it up again. This command I have received from My Father" (John 10:18).

"Yet it pleased the Lord to bruise Him; He has put Him to grief. When You make His soul an offering for sin, He shall see His seed, He shall prolong His days, and the pleasure of the Lord shall prosper in His hand" (Isa 53:10).

The celestial point of the spear directed at this sacrificial animal by the Centaur could prophetically correspond to the spear the soldier used to pierce the side of Yeshua following his death on the cross:

> But when they came to Jesus and saw that He was already dead, they did not break His legs. But one of the soldiers pierced His side with a spear, and immediately blood and water came out. And he who has seen has testified, and his testimony is true; and he knows that he is telling the truth, so that you may believe. For these things were done that the Scripture should be fulfilled, "Not one of His bones shall be broken." And again, another Scripture says, "They shall look on Him whom they pierced." (John 19:33–37)

But did you know the blood and water "coming out of" Yeshua's pierced side represents the forgiveness of sins and life-giving, cleansing properties of water?

The constellation Lupus clearly denotes a celestial pictogram of the supreme sacrifice of Yeshua, the Lamb (Ram) of God.

When we look into the night sky (in the southern hemisphere), may we observe with greater appreciation this noteworthy constellation as God's signature in the stars.

Constellation Name	Translation
Lupus (Latin)	wolf (the victim)
Thēríon (Greek)	a wild animal

Enclosed are derivatives and similar words in ancient languages which help clarify the core meaning of this constellation/star name.

Ancient Hebrew/ Aramaic	Phonetic Spelling	Summarized Meaning
Hebrew בְּהֵמָה	bᵉhêmâh	beast, cattle, animal, beasts (coll. of all animals) cattle, livestock (of domestic animals) wild beasts

Ancient Gentile Language	Phonetic Spelling	Summarized Meaning
Latin	lepus	wolf
Greek θήρα	thēra	a wild animal (Rom 11:9)
Greek θηρίον	thēríon	an animal, a wild animal, any animal
Greek θυσία	thysía	sacrifice (the act or the victim, literally or figuratively), sacrifice
Persian	sabu	a beast of prey, a wild beast
Arabic	bahīma	animal
Akkadian	būlu	animals, livestock
Egyptian Arabic	biheem	farm animal
Persian	bahīma	an animal wild or tame

Scripture References

Matt 7:15; Judg 5:27; Ps 137:8

8

The Constellation: Corona

THE NAME *CORONA* IS the Latin word for "crown." The Hebrew word for Corona is Atarah, or "crown." The word *coronation*, meaning "the ceremony of crowning a sovereign or a sovereign's consort," comes from the Latin *corona*.

The constellation Corona is fashioned in the shape of a crown by God, our Creator. It is a symbol of everlasting dominion and authority, "dominion" meaning stewardship or sovereignty over a kingdom while possessing the authority to enforce its laws. Moreover, Corona exemplifies the original preeminence God gave to Adam and Eve over his earthly creation: "So, God created man in His own image; in the image of God He created him; male and female He created them. Then God blessed them, and God said to them, 'Be fruitful and multiply; fill the earth and subdue it; have dominion over the fish of the sea, over the birds of the air, and over every living thing that moves on the earth'" (Gen 1:27, 28).

By their act of disobedience to God in the garden of Eden, Adam and Eve bowed their knee to a foreign god. They became separated from God, their Creator, and lost their "crown" and dominion. Their legal dominion, as represented by Corona, was transferred to Satan, God's enemy, who became (and is) the god of this world (2 Cor 4:3,4).

God's authority reigns supreme over creation. God's authority and the authority of his Messiah can never successfully be contested by the enemy. Yet, only God can intervene and restore the delegated authority he originally bestowed upon Adam and Eve.

In the night sky, the constellation Ophiuchus, known as "the Serpent Restrainer," is adjacent to Corona. Ophiuchus is pictured restraining Serpens, the long snake winding its way upward in an attempt to steal Corona, the crown.

The constellation Ophiuchus represents the Messiah's successful struggle over the forces of the evil one. The constellation Serpens represents Satan, the arch enemy of God, who rebelled and interfered with God's creation. Satan's goal was to steal the dominion and authority God gave Adam and Eve. Therefore, Satan devised a scheme to tempt man to sin and lose his dominion.

When Yeshua died on the cross, however, he spiritually and legally crushed the headship, or authority, of Satan. Hence, Yeshua rose from the dead with all power and authority. Corona the "crown" is worn symbolically, as illustrated in the constellation Cepheus—a picture of the reigning Messiah, the King of kings, and King of the universe.

Everything Adam lost in the fall, Yeshua, the "last Adam," regained—including dominion's crown and more. Everything in heaven and on Earth is based on authority. God's government is based on righteous authority.

A good example of heaven's righteous authority versus man's authority can be found in the book of Matthew:

> Now when He [Yeshua] came into the temple, the chief priests and the elders of the people confronted him as He was teaching, and said, "By what authority are You doing these things?

And who gave You this authority?" But Jesus answered and said
to them, "I also will ask you one thing, which if you tell Me, I
likewise will tell you by what authority I do these things: The
baptism of John—where was it from? From heaven or of men?"
(Matt 21:23–25).

According to this passage of Scripture, it is of the utmost importance to
contemplate our deeds. We must ask ourselves if we are living in accordance with heaven's revealed authority given in the word of God.

The heavenly constellation Corona symbolically fits the description
of the mystery of the seven stars, as referenced by the apostle John in the
first chapter of the book of Revelation.

Much of the symbolic imagery revealed to the apostle John in the
book of Revelation is multidimensional. This symbolic imagery is recorded in Scripture, confirmed in the stars, and shown to John by the
angel of the Lord. In order to understand the symbolic astronomical language penned by John in the first chapter of the book of Revelation, I am
proposing the following possible interpretations of Rev 1:12–20:

Verse 12: "Then I turned to see the voice that spoke with me. And
having turned, I saw seven golden lampstands . . ."

*These seven golden lampstands may correspond to the seven stars of
Ursa Major (Dover), the Greater Sheepfold. In biblical astronomy, Ursa
Major (the Greater Sheepfold) represents the church and body of Christ.
Ursa Major points to Arcturus in Boötes, the Shepherd. Boötes represents
Yeshua, the Good Shepherd, and the geographical seven churches of Asia.
The shape of the constellation Boötes is similar to a first-century Jewish oil
lamp symbolizing God's people as "light" in a dark world.*

Verse 13: "and in the midst of the seven lampstands One like the
Son of Man, clothed with a garment down to the feet and girded about
the chest with a golden band."

*This corresponds to the constellation Boötes, representing Yeshua's
dual role as Priest and Shepherd.*

Verse 14: "His head and hair were white like wool, as white as
snow . . ."

*This verse corresponds to the constellation Aries, symbolic of the white
wool of the spotless, perfect Lamb of God.*

Verse 14 (cont'd): "and His eyes like a flame of fire . . ."

*This verse may correspond to the Pleiades which, in the ancient world,
were called "the eyes of heaven."*

Verse 15: "His feet were like fine brass, as if refined in a furnace . . ."

This verse may correspond to the constellation Perseus. Following the crucifixion of the Messiah on Wednesday evening 4/25/AD31 at 6:00 p.m., the constellation Perseus, the Warrior began descending beneath the western horizon (the Sun aglow at his feet), symbolically holding Satan's head (star: Al Gol) in his left hand. The name Perseus means "breaker" in Hebrew. The star Al Gol is an abbreviation of "Golgotha," the Hebrew word for skull, or head. Following the resurrection of the Messiah on Sunday morning at 4:30 a.m., the constellation Perseus arose (the Sun aglow at his feet), just before sunrise. This imagery indicates that the headship authority of Satan has been eternally broken and removed by the Messiah. Yeshua has indeed conquered Satan and death! According to biblical imagery, "brass" is a symbol of judgment.

This scripture illustrates God's judgment on Satan, crushing "the prince of the power of the air" and the future judgment of the wicked. (See Ezek 1:7, "and they [the feet of the living creatures] sparkled like the color of burnished brass.")

The following scriptures explain the severity of God's judgment of Satan:

"Now is the judgment of this world; now the ruler of this world will be cast out. And I, if I am lifted up from the earth, will draw all peoples to Myself" (John 12:31, 32).

"He who sins is of the devil, for the devil has sinned from the beginning. For this purpose, the Son of God was manifested that He might destroy the works of the devil. Whoever has been born of God does not sin, for His seed remains in him; and he cannot sin, because he has been born of God" (1 John 3:8, 9).

Verse 15 (cont'd): "and His voice is like the sound of many waters . . ."

This may correspond to Aquarius, the water bearer, pouring out living water and blessings on his people. This description is found in Ezek 43:2; Rev 14:2, 19:6; Ps 93:4; Isa 17:13.

Verse 16: "He had in His right hand seven stars . . ."

This corresponds to the seven-star constellation Corona, the Crown. Corona represents the dominion Adam lost in the fall, that was regained by the Messiah, who is called the last Adam.

Verse 16 (cont'd): "out of His mouth went a sharp two-edged sword . . ."

This may correspond to the constellation Pegasus, which symbolizes the heralding of glad tidings. The inspired (God-breathed) word of God brings revelation knowledge to the hearer. This two-fold revelation is likened unto a sharp two-edged sword. The sword symbolizes separation, either separation to God or separation from God. Psalm 19 plays a crucial role in understanding the two-fold revelation by which the Creator, Redeemer reveals himself to mankind: the first revelation is given through the record of astronomy; the second revelation is brought forth through the record of Scripture. These two methods of revelation typify the "two-edged sword." The psalmist, David, states, "The heavens declare the glory of God" (Ps 19:1). The Bible teaches that one is enabled to come to faith in God through witnessing his celestial message—his signature in the stars. The apostle Paul says: "¹⁷ So then faith comes by hearing, and hearing by the word of God. ¹⁸ But I say, have they not heard? Yes indeed: 'Their sound has gone out to all the earth, and their words to the ends of the world'" (Rom 10:17, 18). Here (verse 18), the apostle is quoting Ps 19:4 regarding the revelation of God given through biblical astronomy. The first edge of the sword is based on the word of God declared from the heavens. Secondly, Ps 19 states "the Torah of the LORD is perfect." This verse refers to the Tanakh, which is an acronym for the three divisions of the Hebrew Bible: the Torah, or teaching, known as the Five Books of Moses; the Nevi'im (Prophets), and the Ketuvim (Writings, hence, TaNaKh). This scriptural revelation includes the writings of the apostles. The second edge of the sword, therefore, represents the word of God declared in the Bible.

Verse 16 (cont'd): "and His countenance was like the sun shining in its strength."

This may correspond to the constellations Aries, the Ram, and Perseus, the Warrior, rising in the east on the morning of Messiah's resurrection just before sunrise, with the Sun aglow beneath their feet disappearing into the sunlight.

Verses 17–18: "And when I saw Him, I fell at His feet as dead. But He laid His right hand on me, saying to me, 'Do not be afraid; I am the First and the Last. I am He who lives, and was dead, and behold, I am alive forevermore. Amen. And I have the keys of Hades and of Death.'"

These "keys" were secured by Yeshua, our Shepherd (represented by the constellation Boötes) through transferred authority.

Verses 19–20: "Write the things which you have seen, and the things which are, and the things which will take place after this. The mystery of the seven stars which you saw in My right hand, and the seven golden lampstands: The seven stars are the angels of the seven churches."

The seven stars of Corona correspond to the angels or pastors of the seven churches of Asia Minor.

The constellations confirm the biblical text of the gospel. A prime example comes from the book of Revelation regarding the seven stars of Corona—denoting the crown of authority.

Yeshua conquered death, hell, and the grave, crushing the Dragon's head, resulting in transferred authority. The headship authority of the Dragon is forever removed from the people of God.

In order to understand the constellation to which John is referring: "having seven stars [at] the right hand of Yeshua," we must consider Boötes' position in the sky—which begs the question: Could the mystery of the seven stars correspond to the constellation Corona? Yes!

"The mystery of the seven stars which you saw in My right hand" (Rev 1:20).

To the right of Boötes, the constellation, Corona glistens in power and authority. The Greek word for "in" is ἐπί/epi and means "on, upon" and the Greek word for "right hand" is δεξιός, ά, όν/dexios and means "on the right hand or right side." The Salkinson-Ginsburg Hebrew New Testament translation also renders (רָאִיתָ בִּימִינִי) "on the right." So, this passage of Scripture can clearly be translated "on the right" or "at the right side." This description pictorially fits the majestic Corona.

Corona is the sign of authority. Comprising seven stars in the shape of a crown, Corona represents dominion lost and dominion found. The dominion Adam *lost* in the fall is *found* through the Messiah. Furthermore, God placed Corona on the right side of the constellation Hercules (*Gibbôr*, in Hebrew). Corona is positioned between Boötes (Messiah, the Shepherd/Priest), and Hercules (Messiah, the Mighty Warrior), revealing Yeshua, the Mighty Warrior, who secured the crown of authority.

In the sky, directly below Corona, the constellation Serpens, the Serpent, illustrates Satan's attempt to steal and retain the crown of authority God gave Adam and Eve at the beginning of creation. Ophiuchus, "the Serpent Restrainer" depicts Yeshua preventing the serpent from eternally possessing Adam and Eve's crown.

The combined constellations—Corona and Boötes—represent the Messiah's governmental reign as Shepherd, Priest, and King.

The book of Revelation expands the meaning to the mystery of the seven stars: "These things says He who holds [κρατέω/krateó] the seven stars in His right hand" (Rev 2:1).

The Greek word hold is κρατέω/krateó, meaning "to be strong, rule." It also means "to use strength, i.e., seize or retain (literally or figuratively)—hold (by, fast), keep, lay hand (hold) on, obtain, retain, take (by)." This is precisely what the Messiah accomplished when he recovered and secured Adam's fallen crown. He secured and provided salvation for the whole world.

The crown is not only (figuratively) at the right hand, but is held securely by King Messiah. In order for Yeshua (Jesus) to lay hold of this crown, he first had to lay down his life for you and me. He conquered death, hell, and the grave to gain the prize. Now he enables us, his people, to reign with him as heirs of his everlasting kingdom. Messiah Yeshua is King of kings and LORD of lords.

An example of restored dominion, righteousness, peace, joy, healing, and well-being for the people of God is found in Ps 103:

> Bless the Lord, O my soul;
> And all that is within me, bless His holy name!
> Bless the Lord, O my soul,
> And forget not all His benefits:
> Who forgives all your iniquities,
> Who heals all your diseases,
> Who redeems your life from destruction,
> Who crowns you with lovingkindness and tender mercies,
> Who satisfies your mouth with good things,
> So that your youth is renewed like the eagle's. (Ps 103:1–5)

The Hebrew word for crown is עֲטַר/atar, meaning "compass, crown. A primitive root; to encircle (for attack or protection); especially to crown (literally or figuratively)—compass, crown."

The constellation Corona shines brightly as a reminder of God's amazing love. God enables us to share in his victory. He enables us to share in his word, redemption, life, love and mercy, renewed vigor, strength, and vision.

Charles Wesley's classic hymn "Amazing Love! How Can It Be" contains this awe-inspiring line: "Bold I approach the eternal throne / And claim the crown, through Christ my own."[1]

After his resurrection, Yeshua remained with his disciples for forty days. On the day of his ascension he reminds us of his supreme authority and the delegated authority he gives his people: "And Jesus came and spoke to them, saying, 'All authority has been given to Me in heaven and

1. Wesley, "Amazing Love."

on earth. Go therefore and make disciples of all the nations, baptizing them in the name of the Father and of the Son and of the Holy Spirit, teaching them to observe all things that I have commanded you; and lo, I am with you always, even to the end of the age.' Amen" (Matt 28:18–20).

On the day Yeshua was taken up, two angels spoke to the first disciples, calling them "Men of Galilee." Even though these men were disciples of Yeshua, they were still of Galilee ("of" denoting origin). This is a practical principle regarding our ministry to others. As ambassadors of heaven, we are to be heavenly minded, so we are of earthly good. We can utilize ancestry, geography, history, archaeology, science, and astronomy to reach our neighbors, relatives, and nations with the gospel of Messiah.

> Now when He [Yeshua] had spoken these things, while they watched, He was taken up, and a cloud received Him out of their sight. And while they looked steadfastly toward heaven as He went up, behold, two men stood by them in white apparel, who also said, "Men of Galilee, why do you stand gazing up into heaven? This same Jesus, who was taken up from you into heaven, will so come in like manner as you saw Him go into heaven." (Acts 1:9–11)

The name of the Lord Jesus, "Yeshua the Messiah," represents his nature and authority. The apostle Paul sums up the lifestyle and authority of the believer: "And whatever you do in word or deed, do all in the name of the Lord Jesus, giving thanks to God the Father through Him" (Col 3:17).

"But now we do not yet see all things put under Him. But we see Yeshua, who was made a little lower than the angels, for the suffering of death crowned [*atarah*] with glory and honor" (Heb 2:8, 9).

Yeshua gained dominion's crown. Can you, through the eyes of faith, see Yeshua—the head and Savior of the church—crowned with glory and honor? He is the First, the Last, the Alpha and the Omega (Greek), the Aleph and the Tav (Hebrew), the A and the Z (English) and everything in between. It is he who lives, who was dead, and who is alive forevermore. He is the one who holds the keys of hell and death. One day death—the last enemy—will be destroyed.

In the interim, he expects every enemy to be put under his feet and under the feet of his people (Heb 10:12–14; Rom 16:20; Eph 1:15–23).

He must reign until he has put all enemies under his feet (Ps 110:1 and 1 Cor 15:25, 26). By means of delegated authority, he has given his people the right to reign as heirs of his kingdom and to share in his victory. Through his high priestly ministry, he enables us to pray with results.

"Most assuredly, I say to you, whatever you ask the Father in My name He will give you. Until now you have asked nothing in My name. Ask, and you will receive, that your joy may be full" (John 16:23, 24).

Through his kingly reign in our lives, Yeshua enables us to speak to "mountains," that they will be removed: "Have faith in God. For assuredly, I say to you, whoever says to this mountain, 'Be removed and be cast into the sea,' and does not doubt in his heart, but believes that those things he says will be done, he will have whatever he says. Therefore, I say to you, whatever things you ask when you pray, believe that you receive them, and you will have them" (Mark 11:23, 24).

In view of the Lord's soon return, during a time when the love of many is growing cold, the prophet Malachi reminds us that the Lord still has a faithful remnant who fears (reveres) him. The following promise from the book of Malachi is reminiscent of the constellation Corona:

> Then those who feared the Lord spoke to one another,
> And the Lord listened and heard them;
> So a book of remembrance was written before Him
> For those who fear the Lord
> And who meditate on His name.
> "They shall be Mine," says the Lord of hosts,
> "On the day that I make them My jewels.
> And I will spare them,
> As a man spares his own son who serves him."
> Then you shall again discern
> Between the righteous and the wicked,
> Between one who serves God
> And one who does not serve Him. (Mal 3:16–18)

Praise Yᵉhôvâh's Messiah! He is gathering jewels for his eternal crown of glory. As heirs of his kingdom, we will someday receive crowns resembling the grandeur of the constellation Corona, and the Messiah's everlasting crown. We will cast our crowns before God's throne in everlasting worship and gratitude (Rev 4:10, 11).

Following the resurrection of the Messiah (April 29, AD 31), Corona could be seen in zenith position precisely at midnight. The heavens truly declare Yeshua is risen with all power and authority!

As we look up to observe the constellation Corona, may we gain a greater appreciation for this glorious sign—God's signature in the stars.

The Seven-Star Crown Constellation, Corona (Photo by Tony and Daphne Hallas, http://www.astrophoto.com.)

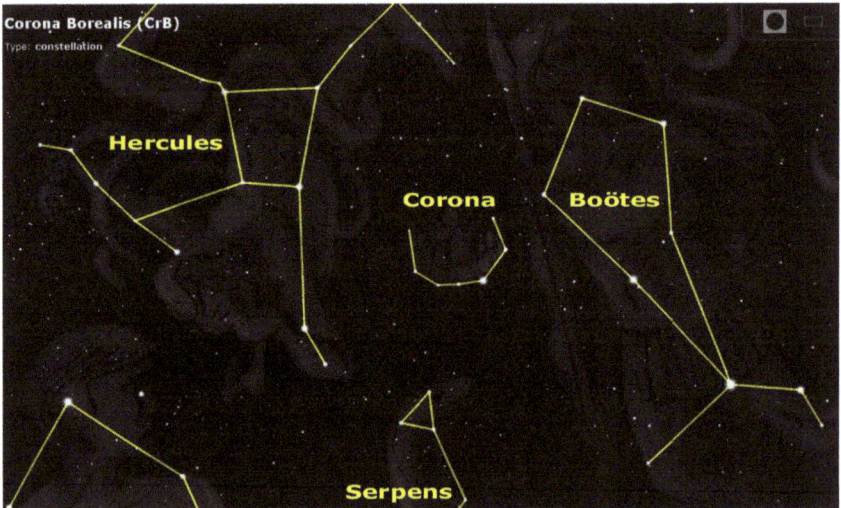

"The mystery of the seven stars which you saw in My right hand" (Rev 1:20). Corona represents the dominion Adam and Eve lost to Satan but is restored by King Messiah. The constellation Boötes represents the Shepherd and high priestly ministry of Yeshua with the crown of dominion at his right side. The constellation Hercules (Gibbôr, in Hebrew) represents Messiah's conquest over the enemy with the crown of dominion at his right side.

Constellation Name	Translation
Corona (Latin)	crown
Stéphanos Bóreios (Greek)	Corona Borealis (northern crown) (a small constellation said to resemble a crown)
Atarah (Hebrew)	crown
Alaklil (Arabic)	corona
Hāla (Arabic)	corona
Al-I'klil Ash-Shamali (Arabic)	the northern crown
běi miǎn zuò (Chinese)	the northern crown constellation
kanmuriza (Japanese)	Corona Borealis, northern crown; a small constellation resembling a crown
Klyl Shmaly (Persian)	Corona Borealis, northern crown; a small constellation said to resemble a crown
AGA (Sumerian)	the crown
A-nim (Akkadian)	the crown

Enclosed are derivatives and similar words in ancient languages which help clarify the core meaning of this constellation/star name.

Ancient Hebrew/ Aramaic	Phonetic Spelling	Summarized Meaning
Hebrew עֲטָרָה	atarah	crown
Aramaic ܐ ܟܠܠܐ	T'aG,aA	crown

Ancient Gentile Language	Phonetic Spelling	Summarized Meaning
Latin	borēus	northern
Latin	Corona	crown
Turkish	kroon	crown
Greek	tiara	crown
Tamil	korōṇā	crown
Arabic	hāla	corona

Ancient Gentile Language	Phonetic Spelling	Summarized Meaning
Arabic	*alaklil*	corona
Arabic	*ṯarā*	wealth
Assyrian	*at ti: ʿru ta*	wealth
Egyptian Arabic	*tharwa*	wealth

Scripture References

Rev 1:20, 2:1; Pss 8:5, 110:1; 1 Cor 15:25–26; Heb 10:12–14; Rom 16:20; Eph 1:15–23; Heb 2:8–9, 1:14; Ps 103:1–5; Matt 21:23–25; Song 3:11; Isa 28:5, 62:3; John 16:23–24; Mark 11: 23–24; Mal 3:16–18

9

The Constellation: Scorpio

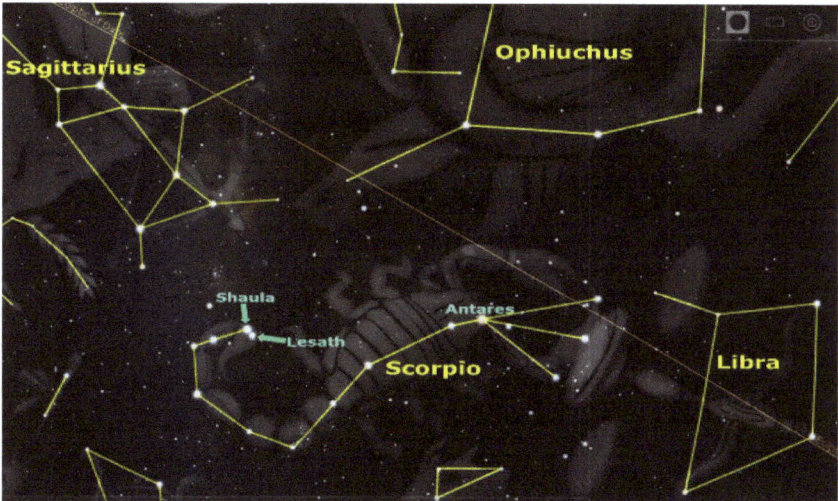

SCORPIO IS THE LATIN word for "scorpion." The Hebrew word for the constellation Scorpio is *Aqĕráb*, or scorpion.

Aqĕráb/עַקְרָב is a derivative of the Hebrew word for war, קְרָב/*qᵉráb*. The Akkadian people named Scorpio "Girtab," "the Seizer," or "Stinger," or "the Place Where One Bows Down"—titles indicative of the creature's dangerous objective.

Scorpio's elongated, semi-circular, tail-like stinger symbolizes a scorpion's *telson*, or "bulb," where its deadly venom is produced and stored. When the Messiah hung on the cross, he took that "sting of death" upon himself as he crushed the serpent's head (authority). With his nail-pierced feet, thrusting himself upward with his heel, Yeshua breathed God's final verdict, "It is finished!" When we receive Yeshua, the Messiah, into our lives, Satan is no longer our god, or "head," of our lives.

Scorpio's claws, or "pincers," are, symbolically, tilting the scales of Libra, revealing Satan's attempt to twist the truth of Scripture and pervert governmental laws using agents of fallen mankind. Scorpio depicts the devil attempting to cause imbalance in our lives. But the Scorpion is crushed by the constellation Ophiuchus, the Serpent Restrainer—a symbol of Messiah.

At the cross, the Messiah crushed the headship authority of the enemy. Therefore, he *will* put an end to injustice. In the interim, Yeshua restored the crown of authority (shown in Corona, the Crown) Satan originally stole from Adam and Eve in the beginning. Satan no longer rules over the people of God. Indeed, Yeshua is the true, sovereign ruler over his redeemed people.

The constellation Sagittarius, the Triumphant Archer, depicts the Messiah shooting his arrows at the heart of Scorpio, our "enemy." The Triumphant Archer represents the Messiah's victory over the kingdom of darkness. Psalm 45:5, the messianic psalm, states: "Your arrows are sharp in the heart of the King's enemies."

Psalm 127 affirms descendants of the righteous: "Like arrows in the hand of a warrior, so are the children of one's youth. Happy is the man who has his quiver full of them; they shall not be ashamed, but shall speak with their enemies in the gate" (Ps 127:4, 5).

The scorpion's claws are composed of (1) Acrab Graffias, *græfias*, Italian for "the claws" (of the scorpion); (2) Dschubba, possibly Arabic for "the forehead"; (3) Nur, Arabic for "light." These three stars illustrate the position of the scorpion's claws. According to R. Hinckley Allen, the origin of these star names may have originated in the Euphrates River area of southwestern Asia.[1] These star-names are not included in E. W. Bullinger's book, *The Witness of the Stars.*

Only through Messiah's delegated authority are we able to put the enemy under our feet. One of the greatest promises to the saints of God regarding our victory over spiritual wickedness is found in the following scripture: "Behold, I give you the authority to trample on serpents and scorpions, and over all the power of the enemy, and nothing shall by any means hurt you. Nevertheless, do not rejoice in this, that the spirits

1. Allen, *Star Names*, 131.

are subject to you, but rather rejoice because your names are written in heaven" (Luke 10:19, 20).

The Israelites of old represent our journey of faith in the Messiah. Similarly, our journey to the "spiritual promised land" is through our own "desert" as we deal with similar tests of faith (2 Pet 1:4; 1 Cor 10:11).

"Who led you through that great and terrible wilderness, in which were fiery serpents and scorpions and thirsty land where there was no water; who brought water for you out of the flinty rock" (Deut 8:15).

Within the scope of Scorpio, the constellation Ophiuchus presents the graphic depiction of God's intervening love for Adam's race. The God of life and the god of death contended for dominion's crown. The god of death was brought down. The God of life is exalted!

God the Creator, through his Son, Yeshua, chose to become a Jewish man in order to redeem mankind. Man was (is) helpless to save himself from the law of sin and death. God knew it would require someone stronger than any natural son of Adam to take back the authority Satan stole from Adam and Eve. God knew it would require the strong right hand of Elohim, the incarnate Son of God, the Son of Man, to defeat the enemy of our souls. Yeshua is depicted in the sky as the strong man taking on the enemy for us. This "strong man" is illustrated by Yeshua in the gospel of Luke: "When a strong man, fully armed, guards his own palace, his goods are in peace. But when a stronger than he comes upon him and overcomes him, he takes from him all his armor in which he trusted, and divides his spoils. He who is not with Me is against Me, and he who does not gather with Me scatters" (Luke 11:21–23). In this illustration, Yeshua portrays the stronger man who overtakes the strong man (Satan) and strips him of his armor and possessions by means of transferred authority.

In the sky, the constellations Hercules (Gibbôr) and Ophiuchus (the Serpent Restrainer) represent Yeshua as the stronger man. In fact, these two constellations are positioned face-to-face as neighboring constellations.

The authority of God, and his Messiah, reigns supreme over creation. God's authority, and the authority of the Messiah, can never been overruled by the enemy. Scorpio, the monster Scorpion, is another graphic depiction of the Messiah's victory over the enemy of our souls. Yeshua, the lover of our soul, has won the war so we can win the battles over lies, deception, and the ways of death. In our quest for truth, we are ever advancing in his strength to enjoy the abundant life Yeshua purchased for his people.

Epic Discovery

On resurrection morning (before sunrise), Sunday, April 29, AD 31, at 4:45 a.m. Ophiuchus, the Serpent Restrainer (representing the Messiah) is pictorially crushing Scorpio, the Scorpion (representing Satan) as it descends beneath the southwestern horizon.

The heavens display comparable positions of the constellations during other proposed dates of the crucifixion and resurrection from AD 27—AD 34. Each year during the biblical spring feasts, God calibrates the sky to reveal these dramatic scenes. The heavens, literally, proclaim Yeshua as victorious over death, hell, and the grave!

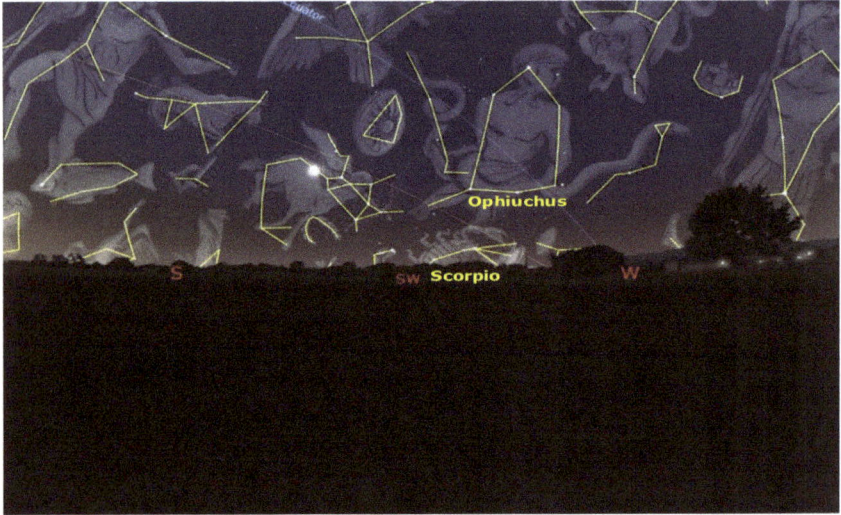

Stellarium image: This astronomical configuration occurred April 29, AD 31, after the resurrection of the Messiah. On Sunday morning, April 29, AD 31 at 4:45 a.m. the constellation Ophiuchus, the Serpent Restrainer (representing the Messiah), is crushing Scorpio, the Scorpion (representing Satan), as it descends beneath the southwestern horizon. There are also several other significant signs occurring here. Yeshua has indeed conquered Satan and death!

Constellation Name	Translation
Scorpio (Latin)	scorpion
Aqĕráb (Hebrew)	scorpion
Skorpios (Greek)	the scorpion
Al-'Aqrab (Arabic)	the scorpion
Aqrab (Persian) Zubāniyāni	the sign Scorpio the two claws (of a scorpion)
GÍR.TAB (Sumerian)	the scorpion
Zuqaqipu (Akkadian)	the scorpion

Enclosed are derivatives and similar words in ancient languages which help clarify the core meaning of this constellation/star name.

Ancient Hebrew/ Aramaic	Phonetic Spelling	Summarized Meaning
Hebrew עַקְרָב	aqĕráb	scorpion
Hebrew עַקְרָב	aqrâb	scorpion
Hebrew קְרָב	qᵉrâb	war
Aramaic ܩܪܒܐ	qaRB'aA	war, battle, fighting
Aramaic ܥܩܪܒܐ	eQaRB,aA	scorpion

Ancient Gentile Language	Phonetic Spelling	Summarized Meaning
Latin	Scorpiō	scorpion
Greek	skorpíos	scorpion
Persian	aqrab	scorpion
Arabic	aqrab	scorpion
Turkish	akrep	scorpion
Egyptian Arabic	aa'rab	scorpion
Akkadian	girtab	the seizer, or stinger, and the place where one bows down

Scripture References

Deut 8:15; Pss 91:13, 45:4–5, 144:1; 2 Pet 1:4; 1 Cor 10:11; Luke 10:19; Rev 9:10

The Stars

Antares

THE STAR NAME ANTARES is defined as "anti-Aries, or against Aries (the ram)." This two-word combination defines the Greek, *anti*, as meaning "against," and the Latin word *ariēs* as "ram, a male sheep, a male lamb" and a "ram lamb." The two words combined mean "against the ram." The name Antares biblically corresponds to the Greek term "antichrist" and the Hebrew "anti-messiah," or "against the Messiah."

The constellation Scorpio gleams brightly in the night sky, depicting the diabolical brilliance of Satan, the enemy. The claws of this Scorpion are symbolically extended as they attempt to tilt the balanced scales of the constellation Libra—twisting and perverting the truth of Scripture and governmental justice through the agents of fallen men who seek to cause imbalance in our lives. But the Scorpion is crushed by Ophiuchus the Serpent Restrainer—who symbolizes the Messiah.

At the same time, Sagittarius, the Archer, is shooting his arrows (the descendants of the righteousness) at Antares, the heart of the Scorpion. The Archer is yet another depiction of the Messiah's triumph over the kingdom of darkness. The great messianic psalm says, "Your arrows are sharp in the heart of the King's enemies" (Ps 45:5.)

Behind Scorpio, the constellation Ara, the Altar, is depicted as having been overturned by the Scorpion. This is a picture of the enemy perverting justice and attempting to overthrow Christianity.

Antiochus IV Epiphanes, the king of Syria, captured Jerusalem in 167 BC and desecrated the temple by sacrificing a pig on an altar dedicated to a pagan god. This act of desecrating the Jewish temple is another "picture" of the enemy perverting and overthrowing aspects of Christianity—and our educational system. The book of Daniel refers to this event as "the Abomination of Desolation" (Dan 12:11).

In an attempt to prohibit Judaism and to "Hellenize" (to become like the Greeks) the Jews, Antiochus banned Jewish religious practices, commanding his military forces to burn all copies of the Torah.[1]

Antiochus, derived from the ancient Greek name "Antíokhos ('Αντίοχος)," is composed of two elements: *antí* (ἀντί) (against, hostile to, opposition, prevention) and *ékhō* (ἔχω) (to have, possess, contain, own, keep, have charge of). Therefore, the name Antiochus (An-*tee*-oh-cus) means "the one who is opposed to others possessing a thing."

The "Antiochus attitude" attempts to inhibit, stifle, and prevent God's people from prospering according to God's promises—to this day! This worldview is propagated through political concepts or ideologies such as socialism, communism, and marxism. God, on the other hand, desires that his saints possess his kingdom and "fill the earth with the knowledge of the Lord as the waters cover the sea" (Hab 2:14; Dan 7:22).

The seed of the serpent, however, uses false religion, power, money and position in an attempt to stifle the work of God and prevent the saints from possessing the kingdom of God. We must be encouraged, in the Messiah, knowing we "do battle" from a place of victory.

Being clear-eyed and pure of heart, we understand that the ongoing war in which we're engaged is between the seed of the serpent and the Seed of the Woman. Satan (portrayed through Scorpio), the master of deception, is attempting to tilt the scales of justice through religion and the traditions of men, thus, nullifying the commandments of God. The enemy uses "agents of Satan" to tilt the scales through various means; fraudulent governmental elections, for example. Satan cannot create anything. He endeavors to twist absolute truth. For example, the enemy attempts to pervert and distort gender identity. This war—the war the enemy is waging against the saints—is based on lies. The saints wage spiritual warfare based on truth. The saints' warfare is not waged against flesh and blood, but against the serpentine powers of darkness.

An example of the word antichrist (ἀντίχριστος/*antíchristos*) is used in the Greek New Testament as follows: "Who is a liar but he who denies that Jesus is the Christ? He is antichrist [ἀντίχριστος/*antíchristos*] who denies the Father and the Son. Whoever denies the Son does not have the Father; he who acknowledges the Son has the Father also" (1 John 2:22, 23).

Antichrist means "against, and opposed to." Antichrist can also be defined as "in the place of" (Christ). The star Antares corresponds to

---

1. Josephus, *Works of Josephus*, "Antiquities of the Jews," 12.5.4.

the spirit of anti-Christ—the heart of the scorpion. The spiritual mindset of Antares has been on Earth throughout history. Yeshua tells us: "Take heed that no one deceives you. For many will come in My name, saying, 'I am the Christ, and will deceive many'" (Matt 24:4, 5). "For false Christs and false prophets will arise" (Matt 24:24).

The apostle Paul informs us: "Let no one deceive you by any means; for that Day will not come unless the falling away comes first, and the man of sin is revealed, the son of perdition, who opposes and exalts himself above all that is called God or that is worshiped, so that he sits as God in the temple of God, showing himself that he is God" (2 Thess 2:3, 4). This anti-Christ spirit is manifested within the framework of false Christianity, not a temple made by hands.

The ancient (tangible) temples of Israel served as a type of the spiritual reality we have in the Messiah. After Yeshua ascended to the right hand of the Father, God's temple became a spiritual temple. If the Orthodox Jews build a *physical* temple used for animal sacrifice on the Temple Mount in Jerusalem, it would not be the temple of God, but "of" man. Second Thessalonians chapter 2 describes man's illegitimate authority over man, regarding religious matters.

In his book, the prophet Daniel foretells the power of the antichrist, describing it as the "little horn power" brought forth out of Rome. In the Bible, a "horn" represents power and authority. Historians estimate untold millions of people were killed during the dark ages who would not submit to religious rule. Many of the true saints of God continued to survive and thrive during this persecution, saints such as the Waldenses, who took refuge in the alps of Northern Italy.

Antares' message (anti-Aries) pertains to the antichrist system of man who are "against the Ram." We understand that the enemy is more interested in "religion," as a means of deception, than any other institution on Earth. This is why God requires all people to turn to him in repentance and faith, to be followers of the real Messiah/Christ helping those enmeshed in false doctrine to come to terms with the balance of truth.

"I was watching; and the same horn was making war against the saints, and prevailing against them, until the Ancient of Days came, and a judgment was made in favor of the saints of the Most High and the time came for the saints to possess the kingdom" (Dan 7:21, 22).

True followers of the Messiah have the opportunity to possess the kingdom of God. Yeshua instructs us: "But seek first the kingdom of

God and His righteousness, and all these things shall be added to you" (Matt 6:33).

The apostle Paul speaks of the outcome of the Messiah's ultimate authority:

> Then comes the end, when He delivers the kingdom to God the Father, when He puts an end to all rule and all authority and power. For He must reign till He has put all enemies under His feet. The last enemy that will be destroyed is death. For "He has put all things under His feet." But when He says "all things are put under Him," it is evident that He who put all things under Him is excepted. Now when all things are made subject to Him, then the Son Himself will also be subject to Him who put all things under Him, that God may be all in all. (1 Cor 15:24–28)

The Messiah is in the process of putting the enemies of God under his feet and under the feet of his saints. He is enabling the heirs of his kingdom to possess his kingdom promises as we learn to reign in life. We must learn to appropriate his authority in our lives, and to tread on the works of serpents and scorpions, as typified by Scorpio and the serpentine constellations. The Messiah won the war so we can win the battles.

> Thus says the LORD to you: "Do not be afraid nor dismayed because of this great multitude, for the battle is not yours, but God's." (2 Chr 20:15)

Conventional Star Name/Meaning	Name/Meaning Confirmed or Corrected
Antares—the wounding	corrected: anti-Aries (against the Lamb)

Enclosed are derivatives and similar words in ancient languages which help clarify the core meaning of this constellation/star name.

Ancient Hebrew/ Aramaic	Phonetic Spelling	Summarized Meanings
Aramaic ܐܢܛܝܟܪܣܛܘܣ	*anty-krystws*	antichrist

Ancient Gentile Language	Phonetic Spelling	Summarized Meanings
Greek ἀντί	*anti*	against, opposite
Greek	*arēn*	lamb
Latin	*ariēs*	a ram
Greek ἀντίχριστος	*antíchristos*	an opponent of the Messiah: antichrist

Scripture References

1 John 2:18, 4:3

Lesath

The star name Lesath (pronounced LAY-soth) is Persian for "scorpion sting." Lesath is positioned in the tail of the scorpion where its poison is released. The Persian *lesath* and the Hebrew לְזוּת/*lazuth*, meaning "perverseness, perverse," may be related words, derived from the same root.

An example of the use of לְזוּת/*lazuth* in the Hebrew scriptures comes from the book of Proverbs: "Put away from you a deceitful mouth, and put perverse [לְזוּת/*lazuth*] lips far from you" (Prov 4:24).

Fallen human beings, by nature, go forth from the womb speaking lies because they are born with the serpentine seed. "The wicked are estranged from the womb; they go astray as soon as they are born, speaking lies. Their poison is like the poison of a serpent" (Ps 58:3, 4). This is why we must be redeemed by God and born from above so we can be restored to the Father of life through God's Son.

Yeshua spoke truth to the religious leaders of his day telling them they were "of" their father, the devil; *of* denoting origin. He likened these religious leaders to "a brood of vipers of the serpentine kind." Lies and hate, mingled with man-made religion, permeated their lives and careers. Satan cannot create anything. He can only pervert. "You are of your father the devil, and the desires of your father you want to do. He was a murderer from the beginning, and does not stand in the truth, because there is no truth in him. When he speaks a lie, he speaks from his own resources, for he is a liar and the father of it" (John 8:44). "Serpents, brood of vipers! How can you escape the condemnation of hell?" (Matt 23:33)

The mouth is connected to the heart. If there is poison in the heart, it will be released through the mouth. Only the Lord can deliver us from the poison of sin issued by the enemy. Only the Lord can give us a new heart and mind.

The seed of the serpent versus the Seed of the Woman is distinguished by speech. We can discern the seed of the serpent by lies and hate. The Seed of the Woman is discerned by truth and love. One of the chief characteristics of the Messiah is grace and truth: "And the Word became flesh and dwelt among us, and we beheld His glory, the glory as of the only begotten of the Father, full of grace and truth" (John 1:14).

The following classic messianic psalm, Ps 45, describes one of the chief characteristics of the Messiah: "*You are fairer than the sons of men; grace is poured upon Your lips*" (Ps 45:2). This grace would be a sign of the Messiah, the illustrious Seed of the Woman.

After Yeshua's baptism and wilderness temptation, he traveled to Galilee, where he taught the true meaning of the scriptures in their synagogues—being glorified by all. There he read from the scroll of Isaiah (concerning himself) "and they were amazed at the gracious words that proceeded out of His mouth." But when Yeshua said, "Today this scripture is fulfilled in your hearing," they were filled with wrath and tried to throw him off a cliff (Luke 4).

Yeshua was filled with the Ruach Hakodesh (the Holy Spirit) and taught with grace and authority. The Messiah, the holy Seed of the Woman, was known by what he said and did. But there was a conflict of kingdoms. This conflict was between purity and perversion; the clash between the kingdom of light and the kingdom of darkness. This "battle" was encountered first by Yeshua in the wilderness temptation and later in the synagogue. Yeshua faced Satan directly in the wilderness. He faced Satan, indirectly, through the religious men in the synagogue. Thus, the conflict continues to this day.

When we are born of God (born from above), born of the incorruptible seed of the word of God, God produces his nature in our hearts and we learn to speak and live like Yeshua. He gives us new "sight" based on heaven's perspective.

In summary, the star Lesath reminds us of the truth of a proverb: "Put away from you a deceitful mouth, and put perverse [לְזוּת/*lazuth*] lips far from you" (Prov 4:24).

Let us treasure this psalm: "The mouth of the righteous speaks wisdom, and his tongue talks of justice. The law of his God is in his heart; none of his steps shall slide" (Ps 37:30, 31).

Conventional Star Name/Meaning	Name/Meaning Confirmed or Corrected
Lesath—the perverse	confirmed

Enclosed are derivatives and similar words in ancient languages which help clarify the core meaning of this constellation/star name.

Ancient Hebrew/ Aramaic	Phonetic Spelling	Summarized Meaning
Hebrew לְזוּת	*lazuth*	perverseness, perverse
Hebrew לזות	LZWT	perverse
Hebrew לֶשַׁע	*lesha*	to break through, puncture, fissure (the name of a location near Sodom and Gomorrah)
Aramaic	LEuOQSeA	sting
Aramaic	lṣyt	up to, towards

Ancient Gentile Language	Phonetic Spelling	Summarized Meaning
Chaldean	*lesha*	perverse
Arabic	*lasa'a*	sting
Persian	*lasa'a*	to sting
Persian	*las'at*	scorpion sting
Amharic	*ebelashe*	break

Scripture References

Prov 2:15, 3:32, 4:24, 14:2; Isa 30:12; Gen 10:19

Shaula

The star name Shaula (pronounced SHOWL-a) is Arabic for "raised tail." Shaula is a derivative of the Hebrew word סָלַל/*saw-lal*, meaning "to exalt (self); reflexively, to oppose."

E. W. Bullinger does not include the star name Shaula in his book *The Witness of the Stars*. The word *saw-lal* is used twelve times in the Hebrew Bible—with good and bad connotations.

The word *shaula*, סָלַל/*sawlal*, is used in the book of Exodus when God sent Moses to speak to Pharaoh: "As yet you exalt [סָלַל/*sawlal*] yourself against My people in that you will not let them go" (Exodus 9:17).

Pharaoh continually resisted God's grace and warnings when Moses told him to let the Lord's covenant people go free from the bondage of Egyptian slavery. Because Pharaoh refused to let the Hebrew people go—even after ten plagues—God allowed Pharaoh and his army to drown at the bottom of the Red Sea.

In the night sky, Shaula—this same raised tail of the scorpion—is depicted as being crushed by Ophiuchus, the Serpent Restrainer. Ophiuchus is depicted crushing the Scorpion, thereby taking death's "sting" for us. This is a picture of Messiah's victory over eternal death on our behalf.

The chorus of my star song "Dominion's Crown" echoes the demise of the enemy and the exaltation of the Messiah:

> Our Deliverer is the serpent holder
> Our Savior, crushed the scorpion
> The foe is brought down
> For the conqueror of death
> Has won dominion's crown

Conventional Star Name/Meaning	Name/Meaning Confirmed or Corrected
Shaula—(not included in E. W. Bullinger's book)	included: *Shaula*—raised tail, to raise, to lift up

Enclosed are derivatives and similar words in ancient languages which help clarify the core meaning of this constellation/star name.

Ancient Hebrew/ Aramaic	Phonetic Spelling	Summarized Meaning
Hebrew סָלַל	*saw-lal*	to exalt (self); reflexively, to oppose
Hebrew סְלָלָה	*sə·lu·lā h*	cast up
Aramaic ܐܠܐ	*'LA, 'eLaA*	raised, arrogant

Ancient Gentile Language	Phonetic Spelling	Summarized Meaning
Assyrian	*sa: li:q*	to raise
Assyrian	*sellu*	lift up, cast up
Arabic	*sāla*	to raise
Arabic	*sāla*	to raise
Persian	*shaulat*	the tail of a scorpion, or the cocked-up part of it
Persian	*shaukat*	one thorn; a sting (of a scorpion)
Persian	*ṣu 'lūl*	a wart

Scripture References

Exod 9:17

10

The Constellation: Serpens

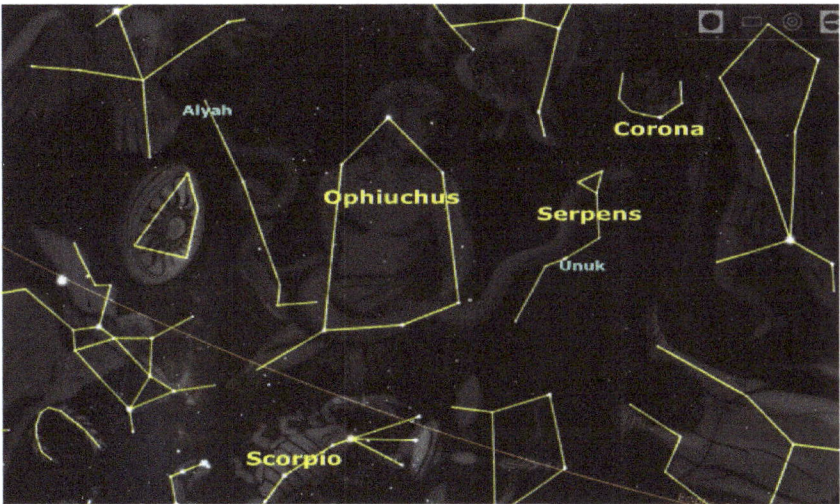

THE NAME *SERPENS* IS the Latin word for "snake." Visible at latitudes between +80° and −80°, the constellation Serpens is best seen during the month of July at approximately 9 p.m.

Serpens, the massive snake, is being held and restrained by Ophiuchus, the Serpent Restrainer. The head of this snake points northward toward Corona, the seven-star crown. Corona, the Crown, represents the dominion God originally bestowed upon Adam and Eve, our first "parents."

Adam and Eve fell into sin and were banished from the garden of Eden by believing the lies of Satan, the enemy. Consequently, Adam's "crown" was (symbolically) transferred to Satan. Thus, Satan became the god of this world (2 Cor 4:4).

Ophiuchus, the Serpent Restrainer, represents the Messiah (the last Adam), who recovers the crown Adam lost in the fall. Because Messiah has recovered this crown of dominion, he delegates his kingdom authority to his people.

The serpent has long since been the symbol of Satan. Originally, the serpent may have had legs and wings. After the serpent allowed Satan to enter its body it became cursed—doomed to crawl on its belly.

"And dust shall be the serpent's food" (Isa 65:25). Man was made of the "dust" of the earth. The devil feeds off fallen humanity. This is why the devil, symbolically, is transformed from a mere serpent to an enormous dragon. He has been consuming "dust" for thousands of years.

Archeological descriptions of winged serpents have been unearthed in ancient Egypt and South America. Yet, the mystery as to how Satan possessed the body of a serpent in the garden of Eden remains. Nevertheless, as Yeshua went about "doing good" during his earthly ministry, demons "came out of the woodwork" through people who were (and are) provoking intense spiritual confrontation.

An example of demon possession transpired in a man known as "Legion." "Legion" describes the multitude of evil spirits he possessed (Mark 5:9). Yeshua cast the demons out of this man allowing these evil spirits to enter a herd of pigs. The pigs committed suicide by running off a cliff, plunging into the sea below and drowning.

Snakes and their venomous poison are symbolically significate in the book of Numbers. When the people complained against God and Moses, the scripture says: "So the Lord sent fiery serpents among the people, and they bit the people; and many of the people of Israel died" (Num 21:6).

After Moses prayed for the people, God instructed Moses: "Then the Lord said to Moses, 'Make a fiery serpent, and set it on a pole; and it shall be that everyone who is bitten, when he looks at it, shall live.' So Moses made a bronze serpent, and put it on a pole; and so it was, if a serpent had bitten anyone, when he looked at the bronze serpent, he lived" (Number 21:8–9).

"Fiery" serpents symbolize the demonic seed of the serpent. "Fiery" most likely refers to the burning pain the afflicted experience when bitten. "Fiery" may also represent the tongues of ungodly men in whom the poisonous serpentine seed is expressed through their words. "They sharpen their tongues like a serpent; the poison of asps is under their lips. Selah" (Ps 140:3; Rom 3:13). Moreover, the fiery serpent on the pole represents the Messiah, who would *be made sin* (a sin sacrifice) for us. The Messiah is the only remedy for sin and sickness.

Land snakes are poisonous whereas the brazen (brass) serpent had no poison. And there was no poison in the Messiah. Just as brass typifies judgment, the Messiah was judged by God for our sins by means of the cross. God also judged Satan, the originator of sin and rebellion. Through the cross, God dethroned the enemy's power over our lives (Col 2:15).

"Now is the judgment of this world; now the ruler of this world will be cast out" (John 12:31).

"And as Moses lifted up the serpent in the wilderness, even so must the Son of Man be lifted up, that whoever believes in Him should not perish but have eternal life" (John 3:14).

The Israelites bitten by the fiery serpents were instructed to look up at the brazen serpent Moses placed on the pole. All who looked (up) were healed. The brazen serpent serves as an illustration of God's provision for our sin today. Today, we are instructed to look to Yeshua and live. Can you look to Yeshua, with eyes of faith, dying on the cross for your sins? Have you looked to the Savior for salvation? Look and live!

The enemy comes to steal, kill and destroy, but Yeshua gives us abundant life (John 10:10). Yeshua came to destroy the works of the enemy (1 John 3:8).

Yeshua is our example of how to tread on serpents and scorpions, and over all the power of the enemy. How did he accomplish this? The Holy Spirit descended upon him in the form of a dove at his baptism (John 1:32). Yeshua proclaimed, "The Spirit of the LORD is upon Me" (Isa 61:1; Luke 4:18). Yeshua submitted himself to the will of the Father (Lk 22:42; Jas 4:7). He spoke to the demons and cast them out with his words (Matt 8:16). He spoke to the mountains and they moved (Mark 11:23). He touched the oppressed and healed them with his anointing (Matt 8:16–17; Acts 10:38). In his name we are commissioned to go and do likewise (Matt 28:19, 20). Demons are subject to us today in Yeshua's name, just as they were to the seventy men Yeshua sent out with the gospel (Luke 10:1–24).

Today, we battle from a position of victory. God is accelerating his kingdom authority. The power of numbers is on the side of the saints.

"How could one chase a thousand, and two put ten thousand to flight" (Deut 32:30). Remember God's angels are allies in the Lord's army: "Do not fear, for those who are with us are more than those who are with them" (2 Kgs 6:16).

Interestingly, the three stars constituting the diamond head of Serpens are pointing away from the crown, as if the heavens are saying, "The Messiah has defeated the enemy, 'turning' him away from the crown as he (the enemy) goes down."

In this regard, astronomy software led me to an intriguing discovery regarding Serpens: Turning back the universe to 4000 BC (the approximate time Adam and Eve were created) while progressively advancing the universe forward in time, we can see the head/mouth area of Serpens opening gradually beneath the constellation, Corona. This may illustrate the enemy's ongoing desire to seize the crown. Our Messiah, however, took that crown from Satan. Now, our Messiah shares his delegated authority with his chosen people—those who choose him.

"Therefore rejoice, O heavens, and you who dwell in them! Woe to the inhabitants of the earth and the sea! For the devil has come down to you, having great wrath, because he knows he has a short time" (Rev 12:12).

"And the dragon was enraged with the woman, and he went to make war with the rest of her offspring, who keep the commandments of God and have the testimony of Jesus Christ" (Rev 12:17).

The constellations Ophiuchus and Serpens portray the fierce contest for the crown of dominion. Yeshua won this contest! As a result, the old serpent/dragon is cast down. Satan's headship authority was crushed at the cross. Therefore, Satan has no authority in the life of God's saints. We must guard our "doors" and not give place to the enemy. Yeshua is reigning at the right hand of majesty with all power and authority, thus enabling us to reign in life with him.

"So the great dragon was cast out, that serpent of old, called the Devil and Satan, who deceives the whole world; he was cast to the earth, and his angels were cast out with him" (Rev 12:9).

Yeshua said to the church of Philadelphia: "Behold, I am coming quickly! Hold fast what you have, that no one may take your crown" (Rev 3:11).

Yeshua has taken hold of the crown of dominion, glory, and victory—illustrated by the celestial crown, Corona—and *we* are to hold fast our crown and not allow men to take it from us.

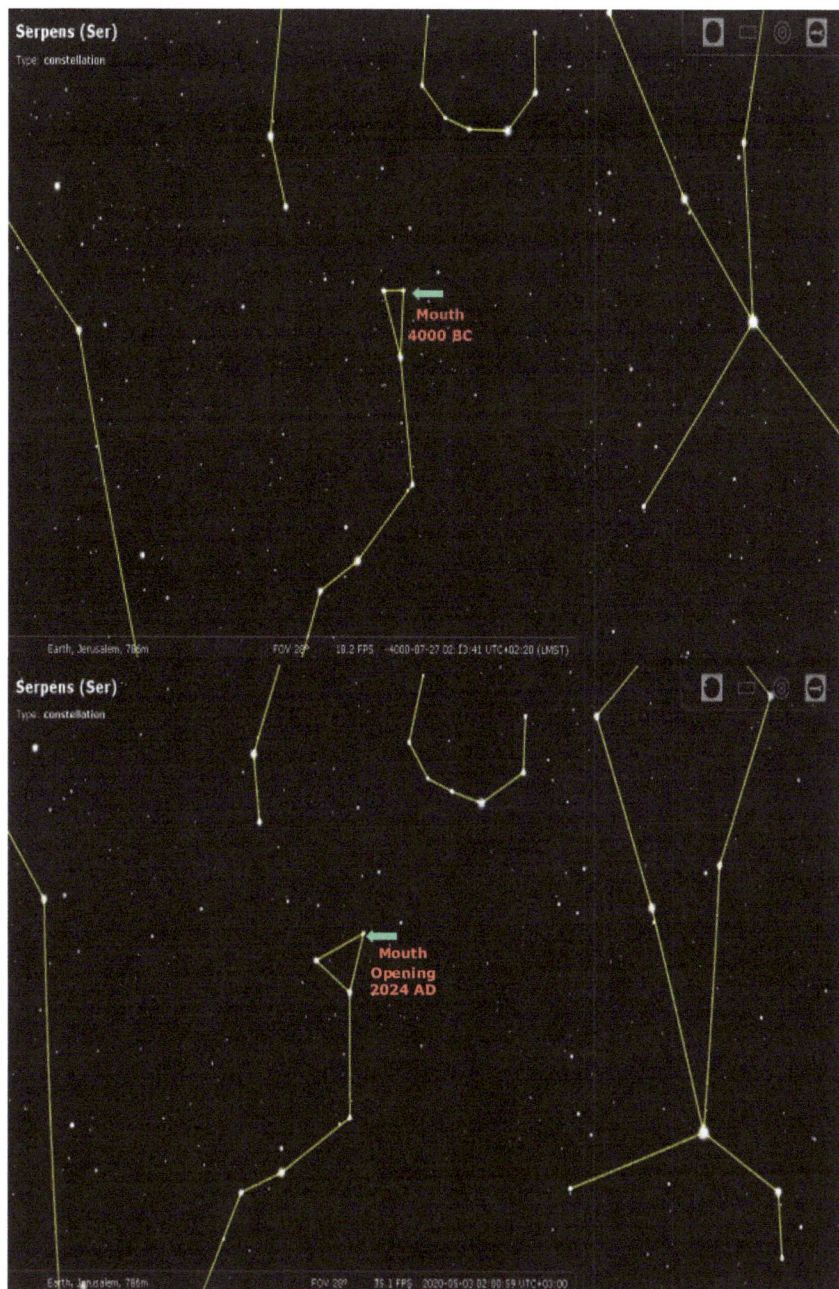

TOP: Stellarium image: 4000 BC—mouth of Serpens partly open.
BOTTOM: Stellarium image: AD 2024—mouth of Serpens wide open.

Constellation Name	Translation
Serpens (Latin)	serpent
Al-Hayyah (Arabic)	the snake
jù shé zuò (Chinese)	the huge snake constellation

Enclosed are derivatives and similar words in ancient languages which help clarify the core meaning of this constellation/star name.

Ancient Hebrew/ Aramaic	Phonetic Spelling	Summarized Meaning
Hebrew שָׂרָף	sârâph	burning, poisonous serpent; specifically, a saraph or symbolical creature (from their copper color):—fiery (serpent, seraph)
Aramaic	srp	venomous

Ancient Gentile Language	Phonetic Spelling	Summarized Meaning
Latin	Serpēns	serpent, snake
Greek ὄφις	óphis	(through the idea of sharpness of vision); a snake, figuratively (as a type of sly cunning) an artful malicious person, especially Satan:—serpent.

Scripture References

Isa 65:25; Num 21:6–9; Gen 3:15; Ps 140:3; Rom 3:13; John 3:14, 12:31; Rev 3:11, 12:9

The Stars

Unuk

THE STAR NAME UNUK is Arabic for the word "neck." Unuk is positioned beneath the neck of the snake. A possible Hebrew word relating to Unuk is חוג/*chûwg*, which means "to encircle, encompass." The word *unuk* is used once in Scripture: "He has compassed [חוג/*chûwg*] the waters with bounds, until the day and night come to an end" (Job 26:10).

Unuk portrays Serpens (the Serpent) encompassing Ophiuchus, the Serpent Restrainer. Serpens represents Satan attempting to retain the crown of dominion, represented by Corona—the crown of dominion Adam lost in the fall. The enemy, however, is restrained. The crown is forever restored by the Messiah.

Conventional Star Name/Meaning	Name/Meaning Confirmed or Corrected
Unuk—encompassing	corrected: neck (of the snake)

Enclosed are derivatives and similar words in ancient languages which help clarify the core meaning of this constellation/star name.

Ancient Hebrew/ Aramaic	Phonetic Spelling	Summarized Meanings
Hebrew חוּג	chûwg	to encircle, encompass
Hebrew עֲנָק	Ânâq	a necklace (as if strangling):—chain from עָנַק/ʻânaq to choke, strangle, encompass

Ancient Gentile Language	Phonetic Spelling	Summarized Meanings
Arabic	unuq	neck
Persian	unuq	neck
Assyrian	khu: ʻga ia	circle

Scripture References

Job 26:10; Gen 3:1; Ps 73:6

Cheleb

The star name Cheleb is likely from the Hebrew חֵלֶב/chêleb, meaning "fat." Cheleb is an unidentifiable star presumably located in the "head" of the massive snake. The "fat head" of the snake may symbolize the serpent's dreadful growth due to consuming Adam's race.

An example of the use of חֵלֶב/chêleb is in the Hebrew scriptures: "The proud have forged a lie against me, but I will keep Your precepts with my whole heart. Their heart is as fat [חֵלֶב/chêleb] as grease, but I delight in Your law" (Ps 119:69, 70).

Thank God for his restraining power and authority; delivering the sons of Adam from the headship authority of the god of this world.

Conventional Star Name/Meaning	Name/Meaning Confirmed or Corrected
Cheleb—the serpent enfolding	corrected: fat (head of the snake)

Enclosed are derivatives and similar words in ancient languages which help clarify the core meaning of this constellation/star name.

Ancient Hebrew/ Aramaic	Phonetic Spelling	Summarized Meaning
Hebrew חֵלֶב	*cheleb*	fat חֵלֶב/*cheleb, kheh'-leb*; or חֵלֶב/*chêleb*; from an unused root meaning to be fat; fat, whether literally or figuratively; hence, the richest or choice part:—× best, fat(-ness), × finest, grease, marrow

Ancient Gentile Language	Phonetic Spelling	Summarized Meaning
Assyrian	*khil ba*	fat

Scripture References

Ps 73:7; Lev 7:23, 16:25; Ps 17:10, 119:69–70

Alyah

The star name Alyah originated in the ancient Hebrew אָלָה/ *'âlâh*, "curse." Alyah is located at the tip of the tail of the serpent. The "tail" may symbolize the serpent's destiny in the lake of fire.

Through God's covenant with Abraham, Isaac, and Jacob, God extends his blessings to Israel and all who receive Messiah, today. God promises to make his redeemed remnant the head and not the tail, blessed to be a blessing to the nations. "And the LORD will make you the head and not the tail; you shall be above only, and not beneath, if you heed the commandments of the LORD your God, which I command you today, and are careful to observe them" (Deut 28:13).

According to the saints' position in the reigning Messiah, God has spiritually seated his people above all the opposing powers of this age. Through our position of covenantal blessing, we are to appropriate God's kingdom rule in our lives and help each other fulfill our destiny in God: "And He put all things under His feet, and gave Him to be head over all things to the church, which is His body, the fullness of Him who fills all in all" (Eph 1:22, 23).

The Bible says God remembers his covenant to a thousand generations (Ps 105:8). God reenforces the covenant he made with Abraham through the new covenant. And his principal remains true: When a person blesses God's people, they are blessed; when a person(s) curses God's people, they are cursed.

"I will make you a great nation; I will bless you and make your name great; and you shall be a blessing. I will bless those who bless you, and I will curse him who curses you; and in you all the families of the earth shall be blessed" (Gen 12:2, 3).

In the final analysis, Yeshua will judge the nations based on how they treated God's people, the physical and spiritual seed of Abraham:

> Then He will also say to those on the left hand, "Depart from Me, you cursed, into the everlasting fire prepared for the devil and his angels: for I was hungry and you gave Me no food; I was thirsty and you gave Me no drink; I was a stranger and you did not take Me in, naked and you did not clothe Me, sick and in prison and you did not visit Me." Then they also will answer Him, saying, "Lord, when did we see You hungry or thirsty or a stranger or naked or sick or in prison, and did not minister to You?" Then He will answer them, saying, "Assuredly, I say to you, inasmuch as you did not do it to one of the least of these, you did not do it to Me." And these will go away into everlasting punishment, but the righteous into eternal life. (Matt 25:41–46)

From the beginning of the fall of Adam in the garden of Eden, the serpent has been a symbol of Satan, the cursed one. The only way out of this dilemma was for the Messiah, "the Blessed One," to come and die for our sins, thus being cursed with our curse so we could be blessed with his blessings. Someday, in the not-too-distant future, the curse will end. God's blessings, however, will continue.

> And there shall be no more curse, but the throne of God and of the Lamb shall be in it, and His servants shall serve Him. (Rev 22:3)

Conventional Star Name/Meaning	Name/Meaning Confirmed or Corrected
Alyah—the accursed	confirmed

Enclosed are derivatives and similar words in ancient languages which help clarify the core meaning of this constellation/star name.

Ancient Hebrew/ Aramaic	Phonetic Spelling	Summarized Meaning
Hebrew אָלָה	*'âlâh*	from a primitive root; properly, to adjure (usually in a bad sense); an imprecation:—curse, cursing, execration, oath, swearing

Ancient Gentile Language	Phonetic Spelling	Summarized Meaning
Arabic	*bahala*	curse

Scripture References

Deut 28:13; Eph 1:22–23; Ps 105:8; Gen 12:2–3; Matt 25:41–46; Rev 22:3; John 10:10; 1 John 3:8; John 1:32; Isa 61:1; Luke 4:18; Jas 4:7; Matt 8:16–17; Mark 11:23; Acts 10:38; Matt 28:19–20; Luke 10:1–24; Deut 32:30; 2 Kgs 6:16; Rev 12:12, 12:17

11

The Constellation: Ophiuchus

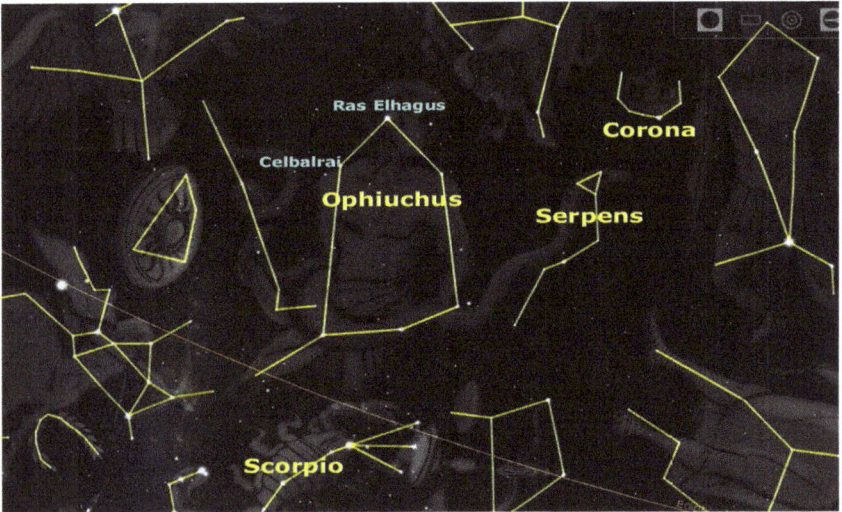

THE NAME *OPHIUCHUS* (PRONOUNCED oh-fee-YOU-kuss) is the title for "Serpent Restrainer." Ophiuchus is a combination of two Greek words: ὄφις/*óphis*, meaning "snake," and κάτοχος/*kátochos*, which means "bearer, holder, restrainer, possessor." *Ophis* is figuratively a type of sly cunning, artful malicious person, as in Satan, the serpent.

The word *ophis* (snake) is found in the book of Luke: "Behold, I give you the authority to trample on serpents [ὄφις/*óphis*] and scorpions, and over all the power of the enemy, and nothing shall by any means hurt you" (Luke 10:19).

A derivative of the word "restrainer," or χάτοχος/*kátochos*, is found in Second Thessalonians: "And now you know what is restraining [χατέχω/*katechó*], that he may be revealed in his own time. For the mystery of lawlessness is already at work; only He who now restrains [χατέχω/*katechó*] will do so until he is taken out of the way. And then the lawless one will be revealed, whom the Lord will consume with the breath of His mouth and destroy with the brightness of His coming" (2 Thess 2:6–8).

Ophiuchus, the Serpent Restrainer, depicts Yeshua, the Warrior/ Shepherd, struggling for the souls of men. Ophiuchus depicts the ancient struggle between good and evil. As the divine Son of God (the Seed of the Woman) was born into the earth, the cosmic conflict intensified. The dragon sought to devour the child through Herod's executioners. The moment the young Messiah rose from the waters of baptism, the enemy was waiting to engage him.

Satan attempted to entice the Messiah to sin during his forty-day fast in the wilderness. His objective was to avert the Messiah's earthly ministry. Yet, the devil was unable to cause Yeshua to stumble. The Scripture says: "Now when the devil had ended every temptation, he departed from Him until an opportune time" (Luke 4:13).

The devil's next strategy was to use religious men to oppose him. After Yeshua ministered in Galilee, where he taught from the scriptures in their synagogues and was glorified by all (Luke 4:15), he traveled to Nazareth where he grew up. There, he read from the scroll of Isaiah concerning himself: "and they were amazed at the gracious words that proceeded out of His mouth" (Luke 4:22). But when he said, "Today this scripture is fulfilled in your hearing," they were filled with wrath and tried to throw him off a cliff (Luke 4:21).

As the hour drew near for Messiah to pay the debt of sin for fallen humanity—a debt he did not owe—the powers of darkness legislated the greatest power-struggle of all time: the battle between everlasting life and eternal death. Messiah Yeshua spent his life-blood conquering the world, the flesh, and the devil. He died and rose from the tomb that he might "save us all from Satan's pow'r when we were gone astray," as stated in that old (circa 1760) hymn "God Rest Ye Merry, Gentlemen."[1]

According to the Scriptures, Yeshua's greatest opposition did not come from the "common" people but from "religious" people—much as it does today. Yet, the salvation of souls is worth the effort!

---

1. "God Rest Ye," Traditional English carol.

As we enforce the victory Yeshua secured, we become more aware of Satan's tactics so we can resist his efforts against us: "Lest Satan should take advantage of us; for we are not ignorant of his devices" (2 Cor 2:11).

The Greek word for *devices*, νόημα/*nóēma*, means "a mental perception, thought, purpose." Although this Greek word is not associated with the constellation name Ophiuchus, its concept is conveyed. Satan, the enemy of our soul, tries to infiltrate our minds with thoughts contrary to the word of God. The enemy's thoughts are intended to bring us down, and to destroy us. God's thoughts and motives are pure and good. They strengthen us to do the task at hand. We must learn to recognize and reject the thoughts of the enemy. We must receive and act on the thoughts God gives us by the Holy Spirit through his written word—the Bible.

The apostle Paul instructs us how to fight in order to win: "For though we walk in the flesh, we do not war according to the flesh. For the weapons of our warfare are not carnal but mighty in God for pulling down of strongholds, casting down *arguments* and every high thing that exalts itself against the knowledge of God, bringing every thought into captivity to the obedience of Christ" (2 Cor 10:3–5). The word *argument* in the Greek language, λογισμός/*logismós*, means "reasoning, conceit, imagination, thought." This struggle takes place on the battlefield of the mind. And where the mind goes, the body follows. What one believes determines how one lives. The enemy is the master of deception. He uses thoughts and ideas—which may seem reasonable—in an attempt to cause us to go astray. "There is a way that seems right to a man, but its end is the way of death" (Prov 14:12).

As illustrated through the constellation Ophiuchus pulling down the powers of the enemy, we must pull down thoughts, arguments, and strongholds contrary to the word and will of God. We can never call a truce with the enemy. We must *stand* in the strength of the Greater One within. "You are of God, little children, and have overcome them, because He who is in you is greater than he who is in the world" (1 John 4:4).

God provides us with defensive armor to wear and the offensive weapon (sword of the Spirit) to bear so we can be successful in this life. His armor is worth its weight, for no weapon can penetrate.

> Therefore, take up the whole armor of God, that you may be able to withstand in the evil day, and having done all, to stand. Stand therefore, having girded your waist with truth, having put on the breastplate of righteousness, and having shod your feet with the preparation of the gospel of peace; above all, taking the

shield of faith with which you will be able to quench all the fiery
darts of the wicked one. And take the helmet of salvation, and
the sword of the Spirit, which is the word of God. (Eph 6:13–17)

We find characteristics of the constellation Ophiuchus in the book
of Jude. The apostle Jude instructs us to "contend earnestly" for the faith:
"Beloved, while I was very diligent to write to you concerning our common
salvation, I found it necessary to write to you exhorting you to contend earnestly for the faith which was once for all delivered to the saints" (Jude 3).

The Greek word for "contend earnestly" is ἐπαγωνίζομαι/
epagōnízomai, which means "to struggle for," or "earnestly contend." This
Greek word is from ἀγωνίζομαι/*agōnízomai* (ag-o-nid'-zom-ahee), the
root of the English term "agonize". *Agōnízomai* in the Greek language
means "to struggle, literally (to compete for a prize), figuratively (to
contend, as with an adversary)," or the possessive case, "to endeavor to
accomplish something; to fight, labor fervently, strive." *Agōnízomai* characterizes the message of Ophiuchus.

Jude instructs us to earnestly contend for the faith which was once
and for all delivered to the saints. Jude's directive includes the foundational context of the true church; the spiritual Israel of God.

Jude does not direct us to earnestly contend for denominational
creeds and traditions. Therefore, we must be diligent in contending for
the true faith.

In the garden of Gethsemane, as Yeshua, the Suffering Servant, prepared to crush the head of the old serpent, he agonized. He agonized with
thoughts of being separated from the Father and nailed to the cross for our
sins. A derivative of the same Greek word, *agōnízomai*, is employed in the
book of Luke regarding Yeshua's struggle: "And being in agony [ἀγωνία/
agōnía], He prayed more earnestly. Then His sweat became like great drops
of blood falling down to the ground" (Luke 22:44). This Greek word is
used exclusively in Luke 22:44, ἀγωνία/*agōnía* (ag-o-nee'-ah), meaning "to
struggle, anguish, agony, a struggle for victory, gymnastic exercise, wrestling of severe mental struggles and emotions, agony, anguish."

We will never know, this side of heaven, the depth of suffering Yeshua, the Suffering Servant, experienced on our behalf. Yeshua was willing to drink from the cup of suffering (plagues) so we could be free from
the curse of sin and death and spend eternity with him.

Contemplating the scenes of the cross helps us gain perspective *in our own struggle* against sin. If we can but fathom the depth of our Father's love for us, we are moved to love him in return.

The Greek translation of the book of Hebrews uses a derivative of *agōnízomai*, to describe our struggle against sin. "For consider Him who endured such hostility from sinners against Himself, lest you become weary and discouraged in your souls. You have not yet resisted to bloodshed, *striving against* [*antagónizomai*: to struggle against] sin" (Heb 12:3, 4).

The Greek word *agōnízomai*, "to struggle," is also used in following passages of Scripture:

"Strive [*agōnízomai*] to enter through the narrow gate, for many, I say to you, will seek to enter and will not be able" (Luke 13:24).

"And everyone who competes for the prize [*agōnízomai*] is temperate in all things. Now they do it to obtain a perishable crown, but we for an imperishable crown" (1 Cor 9:25).

"Fight [*agōnízomai*] the good fight [*agón*, "contest"] of faith, lay hold on eternal life, to which you were also called and have confessed the good confession in the presence of many witnesses" (1 Tim 6:12). The Greek word for "lay hold" is ἐπιλαμβάνομαι/*epilambánomai*, meaning "to seize, to lay hold of, take possession of, overtake, attain, attain to."

Just as Ophiuchus is observed preventing the enemy from retaining the crown, we are, as believers in Messiah, to take hold of the one who gained the crown—Yeshua, our Messiah—and lay hold of eternal life in him.

Psalm 132 describes the authority of the Messiah and our participation with him in his kingdom rule: "His enemies I will clothe with shame, but upon Himself His crown shall flourish" (Ps 132:18).

May we wisely employ God's strength in this struggle for supremacy. Yes, the war has been won, yet battles and skirmishes continue. Sin and Satan may challenge us, but we win—in God's wisdom and strength. In the Messiah we fight the good fight of faith. In the Lord of Hosts, we're fighting a winning battle. "Now thanks be to God who always leads us in triumph in Christ, and through us diffuses the fragrance of His knowledge in every place" (2 Cor 2:14).

Let us remember, we are born into a fallen world, a death zone, a war zone. This battle has been raging from generation to generation. May we choose to continually seek and discover God's purpose for our lives. "For without a vision, the people perish" (Prov 29:18). Truth overcomes lies. "Then Jesus said to those Jews who believed Him, 'If you abide in My word, you are My disciples indeed. And you shall know the truth, and the truth shall make you free'" (John 8:31, 32).

Yet, it's not enough to "read" the Bible. As disciples of Yeshua, we are to *live* by every word of God. The more truth we know and apply the greater freedom we experience. May we learn to embrace God's liberating truth in every area of our lives.

The ancient cosmic conflict portrayed in the sky has come to the earth and found its fulfillment in the Messiah—the conqueror. The Lord of Hosts is enlisting his soldiers to fight the good fight of faith. Will you be his soldier?

The constellation Ophiuchus portrays the Messiah as engaged in a match of strengths. He was (metaphorically) encircled by the serpent's great coils as he prevented the serpent (the enemy of our souls) from retaining the crown he so desperately attempted to hold. At the time and place of God's choosing, the Messiah cast him down by crushing Satan's "head" of authority at the cross.

According to the Scripture, Yeshua existed outside of space and time, and entered our world when the fullness of time had come. He had to die (according to the Scriptures) on Passover, as the Passover Lamb. At his death, he disarmed the enemy by crushing the head of Satan beneath his feet—as portrayed by Ophiuchus crushing Scorpio, the Scorpion. During this act, the heel of the Messiah was symbolically wounded by the sting of the scorpion. This achievement fulfills the first messianic prophecy recorded in the book of Genesis. "And I will put enmity between you [the serpent/ Satan] and the woman, and between your seed and her Seed [the Messiah]; He shall bruise your head, and you shall bruise His heel" (Gen 3:15).

The symbolic sting of the scorpion resulted in the physical death of the Messiah but it could not kill his spirit. Yeshua's final words on the cross were, "Father, into Your hands I commit My spirit" (Luke 23:46).

Yeshua victoriously arose with all power and authority, having the keys of Hades and death. "All authority has been given to Me in heaven and on earth" (Matt 28:18).

"I am He who lives, and was dead, and behold, I am alive forevermore. Amen. And I have the keys of Hades and of Death" (Rev 1:18).

Yeshua restored the crown Adam and Eve lost to the devil in the garden of Eden. This crown is depicted as the constellation Corona. Yeshua restores all things and grants his delegated authority to his people on Earth. Let us not be cheated out of our inheritance in the Messiah. Let us not be taken out prematurely as a casualty of spiritual war. Let us fulfill our destiny in the Messiah and finish our course!

Let us be of good cheer, knowing the Messiah has conquered the powers of darkness and grants us his authority. He writes our names in the Lamb's Book of Life.

"Behold, I give you the authority to trample on serpents and scorpions, and over all the power of the enemy, and nothing shall by any means hurt you. Nevertheless, do not rejoice in this, that the spirits are subject to you, but rather rejoice because your names are written in heaven" (Luke 10:19, 20).

Yeshua overcame the enemy, thereby enabling his people to follow in his footsteps and overcome. "These things I have spoken to you, that in Me you may have peace. In the world you will have tribulation; but be of good cheer, I have overcome the world" (John 16:33).

Yeshua abolished death for all time: "But has now been revealed by the appearing of our Savior Jesus Christ, who has abolished death and brought life and immortality to light through the gospel" (1 Tim 1:10).

The Greek word for abolished is καταργέω/katargeó, which means "to render inoperative." Because the Messiah has abolished death, we have the opportunity to reign in newness of life. Through our new identification in the Messiah, we are able to reckon (count it to be so) our old nature crucified (dead and gone), and our new nature flourishing in him.

The old song "I'll Fly Away" has the line "when I die, Hallelujah, by and by, I'll fly away."1 Considering the day in which we live, we might sing—"if I die," because when Yeshua returns, there will be a generation of saints living on earth who will not pass through the grave but will meet him in the air, transformed into immortality. Halleluyah![2]

In view of the constellation Ophiuchus, let us remember these encouraging scriptures regarding the Messiah's victory over death, the last enemy:

> Precious in the sight of the LORD is the death of His saints. (Ps 116:15)

> For to me [Paul], to live is Christ, and to die is gain. (Phil 1:21)

> He will swallow up death forever, and the Lord God will wipe away tears from all faces. (Isa 25:8)

2. Brumley, "I'll Fly Away."

Your dead shall live; together with my dead body they shall arise.
Awake and sing, you who dwell in dust;
For your dew is like the dew of herbs,
And the earth shall cast out the dead. (Isa 26:19)

Then comes the end, when He delivers the kingdom to God
the Father, when He puts an end to all rule and all authority
and power. For He must reign till He has put all enemies under
His feet. The last enemy that will be destroyed is death. (1 Cor
15:24–26)

All the enemies of God belong beneath his feet, including sickness
and disease, which attempts to attach itself to the saints in Messiah. Writing this chapter on Ophiuchus during the COVID-19 plague reminds
me that sickness is a curse that belongs under our feet. Let us enforce
the Messiah's authority in us, by agreeing with God's promises. Let us
take authority over evil and keep our "immune system" (spirit, soul, and
body) strong in the Lord.

"So, when this corruptible has put on incorruption, and this mortal
has put on immortality, then shall be brought to pass the saying that is
written: 'Death is swallowed up in victory.' 'O Death, where is your sting?
O Hades, where is your victory?'" (1 Cor 15:54, 55).

The constellation Ophiuchus is yet another prophetic picture of the
ancient prophecy (Gen 3:15) of the Messiah. It's the Messiah who destroys all serpentine powers and puts an end to the seed of the serpent.

The stanza in one of my star songs titled "Dominion's Crown" says:
"He was bruised on the heel so we might be healed. He took the *sting* of
sickness so we might live in wholeness . . ."

In an interview with Dr. James Dobson on his popular radio show,
Focus on the Family, author Frank Peretti tells the following story regarding the "sting" of death:

Now, here's a family on vacation. They're driving in their car.
Sunny day, windows are rolled down, breeze is blowing in the
car. They're having a good time. This big old black bee comes
in the window, starts buzzing around in the car. Bzzzzzz. The
little girl sitting in the back seat, she's allergic to bee stings. If
she gets stung, she could die within an hour. "*Help, daddy. It's a
bee. It's going to sting me.*" The father, he's trying to pull the car
over. He's trying to stop. He's trying to catch the bee. He comes
around. He gets up against the windshield. Finally, he catches it,
and gets the bee in his fist. It's in there, and he hangs on, and he

waits for the inevitable. Finally, it happens. Then he lets the bee go. *"Daddy, daddy, it's going to sting me."* "No, honey. He's not going to sting you now. Look what I have in my hand." The bee's stinger is lodged in his hand.

Peretti continues:

> Look what Jesus has in his hand! Satan's sting, the sting of death, the sting of sin, the sting of degradation, the sting of defeat. Jesus took it all. It's in his hand. When you see that nail scar, realize he took it all. He paid it off. He reduced Satan to a big black bee, and all Satan can do is buzz. That's the victory Jesus won for you. Hallelujah, hallelujah. Praise Jesus.[3]

Frank Peretti's story presents an excellent illustration of the constellation Ophiuchus. Jesus took the sting of death for us. He addresses us saying: "'Behold My hands and My feet, that it is I, Myself. Handle Me and see, for a spirit does not have flesh and bones as you see I have.' When He had said this, He showed them His hands and His feet" (Luke 24:39, 40).

> Thanks be to God who gives us the victory through our Lord Jesus Christ! (1 Cor 15:57)

Epic Discovery

On resurrection morning (before sunrise), Sunday, April 29, AD 31, at 4:30 a.m. Ophiuchus, the Serpent Restrainer (representing the Messiah) is pictorially crushing Scorpio, the Scorpion (representing Satan) as it descends beneath the southwestern horizon. At 4:45 a.m. the Scorpion disappears precisely at sunrise.

On Sunday morning, April 29, AD 31, at 4:30 a.m., the constellation Aries (the Ram) rises in the east just before sunrise—the Sun aglow at his feet. The celestial image of the risen Lamb appears (real-time) in the sky confirming his resurrection—God's signature in the stars!

Due to the precession of the equinox (the slow change in the direction of Earth's rotational axis) Aries can no longer be seen rising from Jerusalem before sunrise during the biblical spring feasts because it is hidden by light of the sun.

On resurrection morning, Sunday, April 29, AD 31, at 4:30 a.m. the constellation Perseus rises (from the dark domain), just before sunrise

---

3. Peretti, "God's Way or My Way."

(pictorially), holding Satan's head (Al Gol) in his left hand. Perseus, in Hebrew, means "breaker." The star, Al Gol, is an apparent abbreviation of (Gulgoleth) the Hebrew word for "skull, head."

The heavens display comparable positions of the constellations during other proposed dates of the crucifixion and resurrection from AD 27–AD 34. Each year during the biblical spring feasts, God calibrates the sky to reveal these signs in the heavens.

The heavens are, literally, proclaiming Yeshua as victorious over death, hell, and the grave!

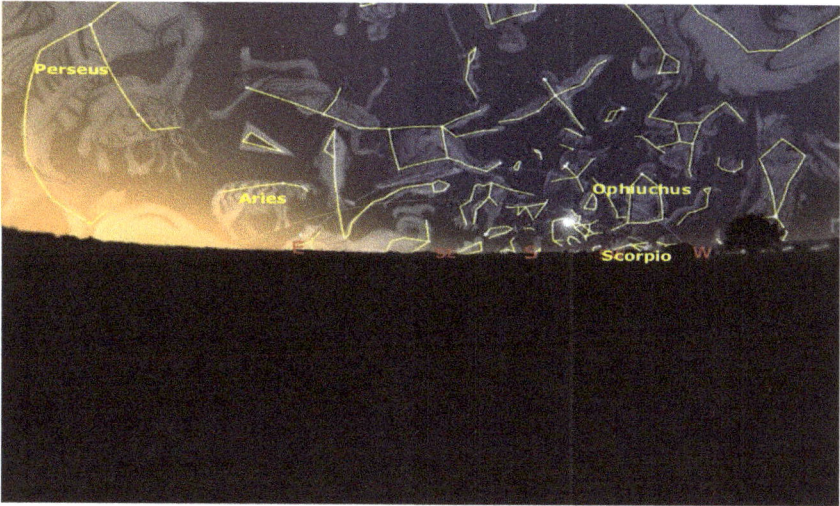

Stellarium image: This astronomical configuration occurred on Sunday morning, April 29, AD 31, at 4:45 a.m. The constellations Aries and Perseus rise just before sunrise depicting Satan's head (Al Gol) in the left hand of Perseus. Perseus in Hebrew means "breaker" and the star *Al Gol* is an apparent abbreviation of (Gulgoleth) the Hebrew word for (skull, head.) This imagery indicates the headship authority of Satan has been utterly broken and removed from the people of God. At the same time Ophiuchus, the Serpent Restrainer (representing the Messiah), is crushing Scorpio, the Scorpion (representing Satan) as it descends beneath the southwestern horizon. Yeshua has indeed conquered Satan and death!

My star song for Ophiuchus, titled "Dominion's Crown," summarizes the story of the gospel.

Dominion's Crown

(Ps 8; Heb 2:5–9; Rev 1:16; 2:1)
By Bruce J. Patterson

Image by Naoyuki Kurita.

Verse 1
In the great struggle
The conflict of the eons
Over the lost dominion
Adam sold out in Eden
Will surely be restored
Completely by our Deliverer

Chorus
Our Deliverer, the serpent holder
Our Savior, crushed the scorpion
The foe is brought down
For the conqueror of death
Has won dominion's crown

Verse 2
He was bruised on the heel
So we might be healed
He took the sting of sickness
So we might live in wholeness
O death, where is your sting
O grave where is your victory

Chorus

Verse 3

Wounded in the contest,
Still, he won the conquest
The head of him who holds
Treading underfoot every foe
The healer of all mankind
See his crown brightly shines—

Chorus

Constellation Name	Translation
Ophiuchus (Greek)	serpent restrainer
shé fū zuò (Chinese)	the snake man constellation

Enclosed are derivatives and similar words in ancient languages which help clarify the core meaning of this constellation/star name.

Ancient Gentile Language	Phonetic Spelling	Summarized Meaning
Greek ὄφις	Óphis	(through the idea of sharpness of vision); a snake, figuratively (as a type of sly cunning), an artful malicious person, especially Satan:—serpent.
Greek κάτοχος	kátochos	bearer, holder, restrainer, possessor

Scripture References

Gen 3:15; Exod 4:4; Luke 10:19–20; 2 Thess 2:6–8; Luke 4:13; 2 Cor 2:11, 10:3–5; Eph 6:12; Prov 14:12; 1 John 4:4; Eph 6:13–17; Jude 3; Rom 11; Prov 11:4; Heb 12:3–4; Luke 22:44, 13:24; 1 Cor 9:25; 1 Tim 6:12;

Ps 134:18; Prov 29:18; John 8:31–32; Matt 5:17–18; Luke 23:46; Matt 28:18; John 16:33; 1 Tim 1:10; Ps 116:15; Phil 1:21; Isa 25:8, 26:19; 1 Cor 15:24–26, 15:54–57; Luke 24:39–40

The Stars

Ras El Hagus

THE STAR RAS EL Hagus is located in the head of Ophiuchus. Ras El Hagus is a combination of three Arabic words: (1) *Ras*, meaning "head," (2) *el*, definite article "the," and (3) Hagus, from *ḥajaza*, meaning "restrain." The star name Ras El Hagus, therefore, means "the head who restrains."

The Hebrew translation for Ras El Hagus is ראש/*rosh*, meaning "head, captain, chief, leader" and אָחַז/*achaz*, meaning "restrains, to grasp, take hold, take possession."

An example of the use of אָחַז/*achaz* in the Hebrew scriptures is demonstrated through the rod of Moses:

> So, the Lord said to him, "What is that in your hand?" He said, "A rod." And He said, "Cast it on the ground." So, he cast it on the ground, and it became a serpent; and Moses fled from it. Then the Lord said to Moses, "Reach out your hand and take [אֱחֹז/ *achaz*] it by the tail" (and he reached out his hand and caught it, and it became a rod in his hand), "that they may believe that the Lord God of their fathers, the God of Abraham, the God of Isaac, and the God of Jacob, has appeared to you." (Exod 4:2–5)

The God of the Hebrews proved to be supreme over the gods of Egypt. The finger of God made a clear distinction between good and evil in overcoming evil with good: "For every man threw down his rod, and they became serpents. But Aaron's rod swallowed up their rods" (Exod 7:12).

This demonstration was a shadow "picture" of the cross—the means by which the Messiah swallowed up the curse. On the cross, Yeshua was cursed with our curse of sin and death. He became the sin-offering to remove our sin. Yeshua, figuratively, took (אֱחֹז/*achaz*) hold of the ancient serpent, pulling him down from his position of authority. The crown of dominion Satan stole from Adam and Eve is now, metaphorically, worn by the Messiah, the King of kings and Lord of lords.

The greater one swallowed the curse on the cross so we could experience his blessing. The greater one lives in his people today so we can overcome the world, the flesh, and the devil.

The star Ras El Hagus, "the head who restrains," represents the headship authority of the Messiah. Yeshua is the head and Savior of his body, the church (the spiritual Israel of God), the bridegroom of his bride. As such, we are to be subject to his headship authority. The Scriptures further teach that Yeshua is the head of all principality and power. This means his headship authority is above every other power, visible and invisible. Yeshua removed the devil's headship authority from the redeemed. The devil's headship authority, however, is still active in the unredeemed world. Only the gospel of the grace of God can save and deliver.

"And He is the head of the body, the church, who is the beginning, the firstborn from the dead, that in all things He may have the preeminence" (Col 1:18)

"And you are complete in Him, who is the head of all principality and power" (Col 2:10).

Recorded in the book of Acts, the story about the apostle Paul who was shipwrecked on the island of Malta is thought provoking. Paul, and the ship's crew, survived a fierce storm at sea and escaped to the island by floating on pieces of the wrecked ship. Arriving on Malta, the native people kindly built a fire for the crew's warmth. Paul gathered a pile of brushwood. As he placed that wood on the fire, a poisonous viper, driven out by the heat, fastened itself to his hand. The Bible says Paul shook it off! This illustrates how we overcome an attack of the enemy. We simply shake it off into the fire from whence it came (Acts 28:1–6).

This story serves as another example of Ras El Hagus, "the head of the restrainer." God's power is over all. God's people ultimately win. As we submit our lives to God, and resist the devil, the Lord is writing the chapters of our lives. Glory to his name!

Conventional Star Name/Meaning	Name/Meaning Confirmed or Corrected
Ras El Hagus—the head of him who holds	confirmed

Enclosed are derivatives and similar words in ancient languages which help clarify the core meaning of this constellation/star name.

Ancient Hebrew/ Aramaic	Phonetic Spelling	Summarized Meanings
Hebrew ראֹשׁ	*rosh*	head, captain, chief, leader
Aramaic	*rish*	head, first
Hebrew אָחַז	*achaz*	restrains, to grasp, take hold, take possession
Hebrew אָחַז	*Âchâz*	he has grasped, possessor, the name of a Jewish king

Ancient Gentile Language	Phonetic Spelling	Summarized Meanings
Greek κατέχω	*katechó*	hold fast, restrain
Arabic	*ḥajaza*	restrains
Arabic	*ḥaqana*	to hold
Arabic	*ra's*	head
Arabic	*El*	definite article "the"
Sumerian	*ha-za*	restrains

Scripture References

Exod 4:35, 7:12; Col 1:18, 2:10; Acts 28:1–6

Mageros

The star name Mageros has its origin in the ancient Greek word μογερός/ *mogerós*, which is defined as "toiling, distressed."

This star name underscores the great toil and distress Yeshua endured to save us all from Satan's power when we were gone astray. We may never know—this side of eternity—what was in the cup of sufferings (plagues) the Savior drank for us all.

The Hebrew word for labor, עָמָל/ 'âmâl, means "toil, trouble, labor." The prophecy of Isaiah says: "He shall see the labor [עָמָל/ 'âmâl] of His soul, and be satisfied. By His knowledge My righteous Servant shall justify many, for He shall bear their iniquities" (Isa 53:11).

The Suffering Servant was willing to endure pain and shame to win us to the Father, as his trophies of grace.

The Scripture says: "who for the joy that was set before Him endured the cross, despising the shame, and sat down at the right hand of the throne of God" (Heb 12:2).

Oh, what a wonderful Savior! We will learn the reward of his labor throughout eternity.

> That in the ages to come He might show the exceeding riches of
> His grace in His kindness toward us in Christ Jesus. (Eph 2:7)

Conventional Star Name/Meaning	Name/Meaning Confirmed or Corrected
Mageros—contending	confirmed

Enclosed are derivatives and similar words in ancient languages which help clarify the core meaning of this constellation/star name.

Ancient Hebrew/ Aramaic	Phonetic Spelling	Summarized Meaning
Hebrew מה	*Mā*	prefix: mā (what)
Aramaic	*grs*	to be made into a snake

Ancient Gentile Language	Phonetic Spelling	Summarized Meaning
Greek μογερός	*mogerós*	of persons, toiling, distressed
Greek γῆρας	*gēras*	old age

Ancient Gentile Language	Phonetic Spelling	Summarized Meaning
Byzantine Greek μα	Ma	prefix: "but"
Ancient Greek μά	Má	prefix: "by"
Persian	qāris	cold; intense, old, ancient

Scripture References

Isa 53:11; Heb 12:2; Eph 2:7

Triophas

The star name Triophas, from the ancient Greek *tropaios*, means "of defeat," from the word *trope* (a rout), originally "a turning" (of the enemy). The word trophy (a prize of war) comes from the Latin *trophæum* (a sign of victory), originally *tropæum*, a transliteration of the Greek *tropaion* "monument of an enemy's defeat." We derive the English word "trophy" from this word. According to the context of this definition, the Messiah has turned the enemy from a place of authority to the place of defeat, from a place of power, to being trampled beneath his feet. The Messiah won the contest and the enemy suffered total defeat. Yeshua won the war so we can win the battles.

The word *trope* is used one time in the Bible: "Every good gift and every perfect gift is from above and cometh down from the Father of lights, with whom is no variableness, neither shadow of turning [*trope*]" (Jas 1:17). According to this scripture, there is no darkness in God. The Father is called the Father of Lights and the Son is called the Light of the World. God's nature is light, and his actions never cast a shadow by turning.

Serpens, the "long snake" constellation, lies directly below Corona and winds through the grasp of the constellation Ophiuchus, "the Serpent Restrainer."

Serpens represents Satan, God's arch enemy. Satan's goal was to steal the dominion God originally gave Adam and Eve. Through his rebellion and subtlety, Satan devised a scheme to tempt Adam (man) to sin, thus losing his God-given authority.

Interestingly, the three stars that make up the diamond head of Serpens is tilted away from the crown, as if the heavens are saying, "The Messiah has defeated the enemy, turning him away from the crown as he goes down."

There are many examples of "a turning" (of the enemy) in the Bible. God turns the enemy away from his people who take hold of his promise: "The Lord will cause your enemies who rise against you to be defeated before your face; they shall come out against you one way and flee before you seven ways" (Deut 28:7).

An example of "a turning" (of the enemy) is found in the story of Esther. Though the name of God is never mentioned in the book of Esther, we see his hand at work behind the scenes. The Jewish people of the city of Shushan were threatened by the villain Haman, a prime minister who convinces King Ahasuerus to kill all the Jews because Mordecai the Jew refused to bow down to Haman. Haman casts "lots" (likely marked stones or jewels) to determine the date upon which the king agrees to destroy the Jewish people. The lots fell on the thirteenth of Adar (March).

In Hebrew, Purim means "lots," and is sometimes called the Feast of Lots. This Jewish holiday commemorates the Lord's act of intervention (during the Persian Empire) which saved the Jewish people from the hands of Haman.

The Jews were saved through the heroic act of Queen Esther—Mordecai's niece, adopted daughter, and wife of King Ahasuerus. Queen Esther came to the kingdom during a time of national danger to stand in the gap for her people, the Jews, and see the hand of God revealed in their lives. Esther's name, Hadassah, means "myrtle tree" in Hebrew. Did you know the flowering myrtle tree emits a fragrance similar to eucalyptus?

When King Ahasuerus discovered his wife Esther's Jewish ancestry, he decided to reverse Haman's decree, thus permitting the Jewish people to defend themselves. Hence, the evil Haman, and his sons, were hanged on the very gallows they built on which to hang Mordecai. God defeated the evil scheme of Haman, and *turned* the tables in favor of the Jewish people. "What then shall we say to these things? If God is for us, who can be against us?" (Rom 8:31).

The ultimate turning point of the enemy's defeat manifested as Yeshua hung on the cross. A great unseen cosmic battle was raging in the heavens and on Earth. In the end, the Messiah triumphed over all the forces of evil—including death and hell. Yeshua did all of this to save us all from Satan's power. Everything Yeshua did, he did for us, in order to make us trophies of his grace.

The cross is the greatest demonstration of the *power* of God to a lost and dying world. It constitutes a turning point of defeat to the enemy—and triumph for the child of God.

"For the message of the cross is foolishness to those who are perishing, but to us who are being saved it is the power [*dunamis*] of God" (1 Cor 1:18). The Greek word for *power* is *dunamis*, meaning "miraculous power, might, strength, miracle working power." Our English word dynamite is derived from the Greek word *dunamis*.

"But you shall receive power [*dunamis*] when the Holy Spirit has come upon you" (Acts 1:8).

Yeshua won the war at the cross. The power of the cross bestows miracle-working power in every area of life. The power of the cross enables us to fight a winning battle in the name of the Messiah.

"Now thanks be to God who always leads us in triumph in Christ, and through us diffuses the fragrance of His knowledge in every place" (2 Cor 2:14).

Through the cross, God turns our sorrow into rejoicing. God turns the universe and the affairs on Earth in our favor. The Lord's love for us never fails.

"And we know that all things work together for good to those who love God, to those who are the called according to His purpose" (Rom 8:28).

"Blessed are the meek, for they shall inherit the earth" (Matt 5:5).

The prophet Daniel was given a vision of God turning the tribulation of the saints into triumph: "I was watching; and the same horn was making war against the saints, and prevailing against them, until the Ancient of Days came, and a judgment was made in favor of the saints of the Most High and the time came for the saints to possess the kingdom" (Dan 7:21, 22).

As previously stated, the star name Triophas is a derivative of *trópaio* from which we derive the English word "trophy." When one is bought and paid for by the precious blood of the Lamb of God, we become trophies of his grace! In the Messiah, we are born to win.

Conventional Star Name/Meaning	Name/Meaning Confirmed or Corrected
Triophas—treading underfoot	corrected: a turning (of the enemy)

Enclosed are derivatives and similar words in ancient languages which help clarify the core meaning of this constellation/star name.

Ancient Hebrew/ Aramaic	Phonetic Spelling	Summarized Meaning
Hebrew רְפַס	*rephas*	to tread, trample
Aramaic ܪܦܣ רְפַס	*raphas*	trampling under foot, stamp

Ancient Gentile Language	Phonetic Spelling	Summarized Meaning
Greek	*trope*	a turning
Ancient Greek	*tropaios*	(*tropaios*, "of defeat"), from trope (a rout) originally "a turning" (of the enemy)
Greek	*trópaio*	prize, trophy
Greek	*pateó*	a path, tread, trample upon from *paíō* (to hit as if by a single blow, especially, to sting (as a scorpion):—smite, strike
Arabic	*rafasa*	to kick

Scripture References

Jas 1:17; Deut 28:7; Esther; 1 Cor 1:18; Acts 1:8; 2 Cor 2:14; Rom 8:28; Matt 5:5; Dan 7:21–22

Celbalrai

The star name Celbalrai is Arabic for "heart of the shepherd."

Similar pronunciations and meanings are found in the selective Gentile languages (see chart).

Celb-al-rai is a combination of three words: (1) celb, from *qalb*, meaning "heart," (2) al, definite article, "the," (3) *rai*, meaning shepherd. The star name Celbalrai therefore means "heart of the Shepherd."

Similarly, the name Caleb is originally a compound word in Hebrew: כֹּל/*kol*, meaning "the whole, all" and לְבָב/*lebab*, meaning "heart, inner man."

The Hebrew counterpart for the Arabic *rai*, "shepherd," is רָעָה/*râ'âh*, meaning "the shepherd, to pasture, to feed." This is the same Hebrew word used for shepherd in Ps 23: "The Lord is my shepherd [*râ'âh*]; I shall not want" (Ps 23:1).

There is nothing more beautiful nor rare than our loving Lord, the Good Shepherd. As described in the book of John, the Good Shepherd gives his life for the sheep and provides pasture for his own. Yeshua calls us by name (John 10:3).

There are several constellations that reveal the "shepherd" characteristics of the Messiah our Shepherd; these include Boötes, Ursa Major and Minor, Auriga, Orion, Cancer, Hercules, and Ophiuchus.

The heart of the Shepherd is revealed in the parable of the lost sheep:

> What do you think? If a man has a hundred sheep, and one of them goes astray, does he not leave the ninety-nine and go to the mountains to seek the one that is straying? And if he should find it, assuredly, I say to you, he rejoices more over that sheep than over the ninety-nine that did not go astray. Even so it is not the will of your Father who is in heaven that one of these little ones should perish. (Matt 18:12–14)

The Good Shepherd cares for each one of us. Each one is of infinite value to the Shepherd—and God is not willing that any should perish.

The star Celbalrai in Ophiuchus (the Serpent Restrainer) serves as a reminder that although Yeshua is a warrior—battling to save us from Satan's power—Yeshua is a shepherd, at heart.

To understand the heart of the Shepherd, we must get a glimpse into the glory of God. "Glory" is an expression of the nature and character of God.

Yeshua is the glory of the only begotten of the Father, full of grace and truth (John 1:14). He is the image of the invisible God (Col 1:15). The heavens declare the glory of God, namely, Yeshua (Ps 19:1).

An expression of the glory of God can be found in the book of Exodus. Moses said to the LORD: "Please, show me Your glory" (Exod 33:18). God responded by having Moses cut out two tables of stone, for the Almighty to write, with his own finger, the Ten Commandments. The commandments of the Lord are an expression of his glory and a transcript of his character. The Lord then said: "I will make all My goodness pass before you, and I will proclaim the name of the LORD before you" (Exod 33:19).

> Now the Lord descended in the cloud and stood with him there, and proclaimed the name of the Lord. And the Lord passed before him and proclaimed, "The Lord, the Lord God, merciful and gracious, longsuffering, and abounding in goodness and truth, keeping mercy for thousands, forgiving iniquity and transgression and sin, by no means clearing the guilty, visiting the iniquity of the fathers upon the children and the children's children to the third and the fourth generation." (Exod 34:5–7)

Below please note the characteristics of the Lord, our Good Shepherd, he exhibited as a man:

- Merciful—compassionate
- Gracious—fair
- Longsuffering—patient, slow to anger
- Abounding in goodness—favor, loving-kindness
- Abounding in truth—assuredly, faithful, right
- Forgiving iniquity, transgression, and sin
- Just—impartial, unbiased, fair

As we consider these fundamental values, we're looking into the shepherd-heart of God. The Almighty is the absolute essence and reality of these values. And the Lord reproduces these characteristics in his people.

The book of Isaiah offers a beautiful passage of scripture expressing the heart of the Shepherd. This passage describes the ministry of Yeshua, as our shepherd (רָעָה/râ'âh): "Behold, the Lord God shall come with a strong hand, and His arm shall rule for Him; behold, His reward is with

Him, and His work before Him. He will feed His flock like a shepherd; [רָעָה/*râ 'âh*] He will gather the lambs with His arm, and carry them in His bosom, and gently lead those who are with young" (Isa 40:10–11).

The prophetic star Celb-al-rai reveals the heart of the Shepherd, who laid down his life for the sheep. The greatest expression of the heart of the Shepherd is sacrifice. Yeshua, our Great Shepherd, was willing to endure the sting of death (symbolized by Scorpio, the Scorpion) so we could be restored to the Father of Life.

Ophiuchus may seem an unlikely constellation in which to discover the heart of the Shepherd. Yet, once we understand its celestial "message of sacrifice" combined with the brilliant star Celb-al-rai, we learn God's message of love has been there all along.

> I am the good shepherd. The good shepherd gives His life for the sheep. (John 10:11)

Conventional Star Name/Meaning	Name/Meaning Confirmed or Corrected
Celbalrai—the serpent enfolding	corrected: heart of the shepherd

Enclosed are derivatives and similar words in ancient languages which help clarify the core meaning of this constellation/star name.

Ancient Hebrew/ Aramaic	Phonetic Spelling	Summarized Meanings
Hebrew לֵבָב	*lêbâb*	heart
Aramaic ܠܒܐ	*LeB'aA*	heart
Hebrew רָעָה	*râ 'âh*	shepherd
Aramaic	*Ra'YaA*	shepherd

Ancient Gentile Language	Phonetic Spelling	Summarized Meanings
Arabic	*qalb*	heart, bosom
Egyptian Arabic	*alb*	heart, bosom

Ancient Gentile Language	Phonetic Spelling	Summarized Meanings
Persian	*qalb*	heart
Turkish	*qalb*	heart
Assyrian	*lib ba*	heart
Phoenician	*lb*	heart
Tigrinya/Ethiopian	*lbi*	heart
Arabic	*ra'ā*	tend
Akkadian	*re'u*	shepherd
Arabic	*rā'i*	wonderful
Persian	*rā'ī*	feeding, grazing; a shepherd, herdsman; a guardian, protector
Assyrian	*ra: 'ia*	sheep, a shepherd
Assyrian	*ra: é*	pasture
Arabic	*ra'a*	shepherd

Scripture References

Ps 23:1; John 10:1–21; Matt 18:12–14; John 1:14; Col 1:15; Ps 91:1; Exod 33:19, 34:5–7; Isa 40:10–11; John 10:11

12

The Constellation: Hercules

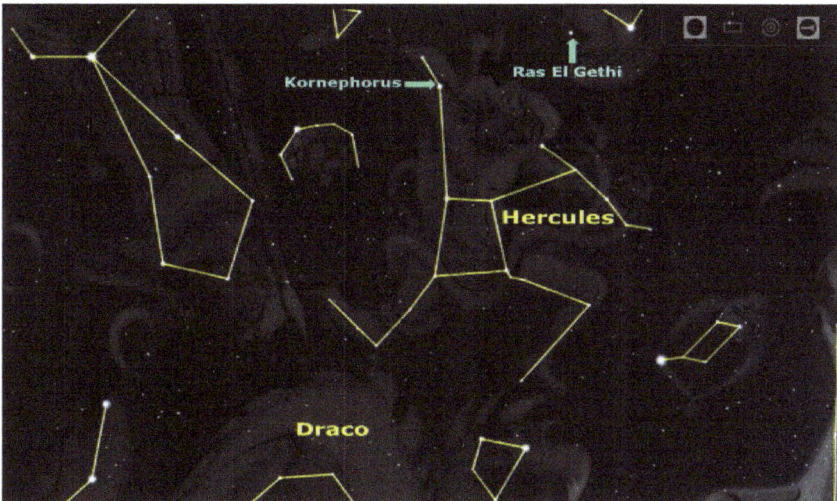

THE NAME HERCULĒS IS derived from the Greek words *hḗrōs*, "hero," and *kléos*, "renown, glory." *Herculēs* is defined as "renown hero." *Gibbôr*, the common or customary Hebrew name for Hercules, is defined as "strong, mighty, hero, and champion."

The constellation Hercules depicts one of the most graphic battles in Messiah's victory over Satan. Through this constellation we can observe the "price" our Creator paid to become our Redeemer. Born of a virgin, the life of the incarnate Son of God would be required as a ransom for the sins of mankind.

During Passover, God—in his mercy—arranged the backdrop of the sky to depict Hercules crushing the head of Draco, the Dragon (Satan). This yearly occurrence reflects the story of the first messianic prophecy given by the Lord in the Bible: "And I will put enmity between you and the woman, and between your seed and her Seed; He shall bruise your head, and you shall bruise His heel" (Gen 3:15). This prophecy was fulfilled by the Messiah Yeshua when he thrust upward on the cross to crush the head of the (invisible) serpent—represented by Draco, the Dragon—thus bruising his heel in the process.

The constellation Hercules illustrates the love of God through pictorial prophecy. As we compare this constellation to John 3:16, we discover God's love for the world; that God, through the Messiah, chose to become a Jewish man in order to redeem mankind.

There are but two "seeds" on Earth: the seed of the serpent (fallen mankind) and the Seed of the Woman (redeemed mankind). Of which "seed" are you? When you choose Yeshua and are born from above, you become part of the redeemed Seed of the Woman (1 John 3:9–10; Eph 2:14, 15).

The constellation Hercules (Gibbôr) is a picture of a Mighty One whose right knee is bending down as if suffering in death. His left foot is over the head of the dragon. This "Kneeling Man" constellation has a star positioned by his head named Ras Al Gethi, which means "head of the kneeler" in Arabic.

Four major events in the Gospels describe Yeshua, the Savior, stooping or kneeling. These four events fulfill the prophetic significance of the "Kneeling Man" as displayed in the sky.

I. Yeshua Meets the Adulterous Woman

Yeshua fulfills the role of the "Kneeling Man" in the Gospel of John. Scribes and Pharisees brought a woman caught in adultery to Jesus. "Jesus *stooped down* and wrote on the ground with His finger, as though He did not hear" (John 8:8). In this verse, we see the Creator of the universe, who formed the heavens and Earth with his fingers, who wrote the Ten Commandments in stone, writing with the finger of God on the ground. A "Torah pointer" is the instrument by which the Jewish people read the Torah. The Torah pointer represents the finger of God. Yeshua *is* the the

Word of God—(Torah) made flesh. The ground represents the dust from which mankind was made (Gen 2:7). This scene represents God writing his commandments in our hearts with his finger (Jer 31:33 and Heb 10:16). It also represents our Creator stooping low in order to give us the right to become "children of God"—new creations in him, enabling us to live godly lives after his likeness (Ps 113:6). We can rejoice in that Satan's headship authority has been crushed beneath our feet by Yeshua, thereby releasing us from the domain of darkness and giving us the right to come into agreement with the word of God.

2. Yeshua, the Foot Washer

The book of John records Yeshua washing the feet of his disciples during the Passover Seder, thus revealing the eternal purpose of God. The Scripture says Yeshua "rose from supper and laid aside His garments" (John 13:3). His gesture represents the laying aside of particular attributes of deity in order to become the "Kneeling Man." "[Yeshua] took a towel and girded Himself. After which He poured water into a basin and began washing the disciples' feet; wiping them with his towel" (John 13:4, 5). This scene characterizes God's Servant coming to serve by washing the disciples' feet in an act of stooping down or kneeling. The Messiah, in effect, became the "Kneeling Man." The washing of the disciples' feet signifies being washed clean from the dirt and grime we acquire as we "walk" through this contaminated world.

Foot-washing is an example of covenantal love. A Christian marriage, for example, is a covenant between two people with God at the center of their relationship. The word "covenant," in the Hebrew language, means "to cut." One who is spiritually bathed in the "covenantal blood of the Lamb" is clean. Yet because we walk through this world—a contaminated environment—we must daily (symbolically) wash our hands and feet with the "washing of the water of the word" (Eph 5:25–26), the Bible, God's love letter to us. This act of "foot-washing" was shown to the disciples to underscore that we are, indeed, our brothers' keeper.

The old covenant priests washed their hands and feet in the laver of brass before entering the holy place of the temple. Brass typifies judgment. Just as water cast the priests' reflection as he gazed into the laver, we are to examine ourselves by looking into the mirror of God's word. We are to see his reflection as we gaze into the "mirror" of God's word. The apostle Paul elaborates on this principle regarding husbands and wives and the church (the spiritual Israel of God). "Husbands, love your wives, just as Christ also loved the church and gave Himself for her, that He might sanctify and cleanse her with the washing of water by the word, that He might present her to Himself a glorious church, not having spot or wrinkle or any such thing, but that she should be holy and without blemish" (Eph 5:25–27).

3. Yeshua, Kneeling in the Garden

Yeshua fulfills the role of the "Kneeling Man" in the garden of Gethsemane as he prepares to crush the Dragon's head: "And He was withdrawn from them about a stone's throw, and He knelt down and prayed, saying, 'Father, if it is Your will, take this cup away from Me; nevertheless, not My will, but Yours be done.' Then an angel appeared to Him from heaven, strengthening Him. And being in agony, He prayed more earnestly. Then His sweat became like great drops of blood falling down to the ground" (Luke 22:41–44). This scripture describes Yeshua "sweating great drops of blood" through his agonizing experience. *Hematohidrosis*, the medical term for this extraordinary occurrence, is a rare disorder in which a human being sweats blood precipitated by extreme stress or exertion. Yeshua foreknew he would experience separation from his heavenly Father through the indescribable agony of becoming the supreme sacrifice for the sins of the world. This reality caused him to sweat great drops of blood, as he was already in the act of crushing the Dragon's head.

4. Yeshua, the Kneeling Man on the Cross of Golgotha

Yeshua fulfills his fourth role as the "Kneeling Man" through tasting death for humanity by crucifixion at Golgotha. Crucifixion engages death by means of suffocation. In order to breathe, the Messiah pushed up on his spike-driven feet, enabling his lungs to fill with air until he no longer had the strength to do so. Suffocating, accompanied by loss of blood from the scourging, with nails driven into his hands and feet triggering extreme mental and physical anguish, the free wall of his heart ruptured causing him to cry out in a loud voice, "It is finished!"

Yeshua *chose* to die on the cross. As his heel was bruised, the invisible head of the ancient serpent (dragon) was crushed. Yeshua's feat of strength fulfilled the Gen 3:15 prophecy and the pictorial prophecy of the constellation Gibbôr. In the final act of crushing the Dragon's head, Yeshua gave up his Spirit to the Father, and bowed his head, his body collapsing into the posture of the "Kneeling Man."

The following verses help us understand the heart of God and why the Messiah was willing to lay down his life for us:

1. "How God anointed Jesus of Nazareth with the Holy Spirit and with power, who went about doing good and healing all who were oppressed by the devil, for God was with Him" (Acts 10:38).

2. Yeshua said, "So ought not this woman, being a daughter of Abraham, whom Satan has bound for eighteen years, be loosed from this bond on the Sabbath?" (Luke 13:16)

The key to understanding how God destroyed the enemy through the cross of Christ is found in a rather obscure scripture: "But we speak the wisdom of God in a mystery, the hidden wisdom which God ordained before the ages for our glory, which none of the rulers of this age knew; for had they known, they would not have crucified the Lord of glory" (1 Cor 2:7, 8).

How could Satan have known exactly how *his headship (authority)* would be crushed according to the ancient prophecy? The arch enemy of God evidently thought—when the Son of God died on the cross and was resurrected—his (Satan's) authority over Adams race would not be overthrown. He could not know his head would be crushed by means of *transferred authority.* He did not know that the Messiah would recover the authority Adam and Eve lost in the fall.

The Jewish rulers and rulers of the Roman government were also ignorant of the gospel revealed in the Hebrew scriptures (see John 12:31, 14:30, 16:11; Acts 3:17, 4:8, 4:26).

During the temptation of the Messiah, when the devil tried to coerce Yeshua to bow to his authority, we read: "Then the devil, taking Him up on a high mountain, showed Him all the kingdoms of the world in a moment of time. And the devil said to Him, 'All of this authority I will give You, and their glory; for this has been *delivered* to me, and I give it to whomever I wish. Therefore, if You will worship before me, all will be Yours'" (Luke

4:5–7). Yeshua did not contest the devil's "authority." He simply rebuked him; directing him to the final authority of God's word.

The Greek word for "delivered," as defined in Luke 4:5–7, is *paradidómi*, "to hand over, to give or deliver over, to betray." *Paradidómi* occurred when man "fell" in the garden of Eden. Adam and Eve, our first parents, surrendered their God-given dominion to Satan. The only way to recover this dominion would be for the sinless Son of God to take back this dominion from Satan by means of his death, burial, and resurrection.

The authority of God reigns supreme over all creation. His authority—the authority of the Messiah—can never be overthrown by the enemy. God intervened, legally, to restore the authority he originally delegated to Adam and Eve. God works his miracles "legally." A legal transaction was enacted through the death, burial and resurrection of the Messiah as he dealt the fatal blow to the head of the serpent dragon. Isn't it amazing to be given the privilege of viewing God's redemptive act portrayed in the night sky through his constellations!

Following the ascension of the Messiah, we find this statement in the book of Revelation: "I am He who lives, and was dead, and behold, I am alive forevermore. Amen. And I have the keys of Hades and of Death. Write the things which you have seen, and the things which are, and the things which will take place after this. The mystery of the seven stars which you saw in My right hand" (Rev 1:18–20).

When the Messiah rose from the dead, he secured the keys of Hades and death. The Greek word for "keys" is κλεῖς, κλειδός, ἡ/*kleis*, defined as "a tool used to secure a lock," literally or figuratively, denoting "power and authority." The Greek word for Hades is Ἅιδης, ου, ὁ/*hadés*, which means "the abode of departed spirits." The Greek word for death is θάνατος, ου, ὁ/ *thanatos*, which means "physical or spiritual death; (figuratively) separation from the life (salvation) of God."

Several things occurred through Yeshua's death, burial, and resurrection. (1) Yeshua paid our sin debt. (2) He came out of the tomb holding the keys of hell and death. (3) The record of our sins was nailed to the cross. (4) Yeshua crushed the Dragon's head resulting in transferred authority (5) Yeshua became the last Adam and the head of the new creation race.

The first two chapters of the book of Revelation reveal the mystery of the seven stars. Did you know this mystery contains one of the greatest truths of all time? Indeed, this mystery reveals the transfer of dominion and authority, which gives new meaning to Luke 21:28: "Now when these things begin to happen, look up and lift up your heads, because your redemption draws near."

The beautiful constellation Corona resembles a seven-star crown. This constellation reveals one of the most important messages of all time: the symbol of God's authority. Corona represents the original dominion God gave to Adam and Eve over his earthly creation. This dominion was stolen by Satan but recovered by the Messiah (please see my chapter on Corona).

"The mystery of the seven stars which you saw in My right hand" (Rev 1:20), corresponds to the constellation Corona positioned between Hercules (Gibbôr), the Mighty God, and Boötes, the Shepherd. In fact, Corona is located, specifically, at the right hand of both constellations.

Note: According to scholarly consensus, the Messiah was crucified on Passover (Nisan 14), Wednesday, April 25, AD 31. He rose from the dead in newness of life, Sunday morning, April 29. Using astronomy software, I discovered amazing astronomical signs in the heavens over Jerusalem when reversing the universe to Wednesday, April 25 and Sunday morning, April 29, AD 31.

Epic Discovery

Did you know every year, during the spring feasts of the Lord, God arranges the sky to "showcase" the constellation Hercules crushing the head of Draco during those hours in which the crucifixion occurred? Could this celestial scene tell the story of the first messianic prophecy (Gen 3:15) given by the Lord in the Bible?

The Bible confirms that Yeshua's crucifixion took place outside the city walls of Jerusalem at "the third hour of the day," or 9:00 a.m. He hung on the cross for seven hours—including the first hour. Yeshua died at 3:00 in the afternoon.

On Wednesday, April 25, AD 31, from 2:00 p.m. to 3:00 p.m. (the hour of the Messiah's death on the cross), the head of Draco, the Dragon (representing Satan, the enemy), descended and disappeared below the northwestern horizon while (symbolically) being crushed by Hercules, the Mighty Gibbôr. During Messiah's crucifixion, the celestial image of this conquering warrior appears (real-time) in the "picture gallery" of the sky, thus confirming the head of the serpent/dragon is crushed by the Savior. This pictorial prophecy illustrates the fatal blow dealt to the authority of the serpent/dragon by the Messiah. This mystery was hidden in God—much like these constellations are hidden by the natural light of the Sun. The heavens are continually announcing that the headship authority of the enemy has been removed from the people of God, as the Mighty Gibbôr crushes the head of the dragon. Heaven's cosmic conflict came to Earth and the Messiah won the war, enabling us to win battles we face.

Scripture verifies darkness covering the earth from 12:00 to 3:00 p.m.: "Now it was about the sixth hour, and there was darkness over all the earth until the ninth hour" (Luke 23:44). The three hours of darkness was a supernatural act of God. A similar phenomenon occurred during the ninth plague on the land of Egypt. For three days the Egyptians were in complete darkness, yet the homes of the Hebrews remained well-lit (Exod 10:21–23).

When Christ hung on the cross, the stars and constellations were possibly visible during those three hours of darkness. This phenomenon would reflect and point to the fulfillment of their heavenly message on Earth—in real-time. As the Messiah laid down his life on our behalf, at 12:00 noon, Cygnus, the Swan, known as the Northern Cross, was descending in the northwest horizon.

The Messiah died upon a "tree," a "cross" assembled from an unknown wood. His was a temporary death as his body laid in a borrowed limestone tomb for three days and three nights: "Whom God raised up, having loosed the pains of death, because it was not possible that He should be held by it" (Acts 2:24 and Ps 16:9–10).

Satan, represented by the serpent/dragon, would experience a *permanent* death blow to his authority, as characterized by the fatal crushing of his head: "And I will put enmity between you and the woman, and between your seed and her Seed; He shall bruise your head, and you shall bruise His heel" (Gen 3:15).

Three days later, on Sunday morning, April 29, AD 31 at 4:30 a.m., the constellation Aries (the Ram) rises in the east just before sunrise—the Sun aglow at his feet. The celestial image of the risen Lamb appears (real-time) in the sky confirming his resurrection—God's signature in the stars!

Just before sunrise, following Messiah's resurrection, the constellation Perseus is rising from the dark domain, holding—as a symbol of victory—Satan's head (Al Gol) in his head in his left hand. Perseus, in

Hebrew, means "breaker." The star Al Gol is an apparent abbreviation of (Gulgoleth) the Hebrew word for "skull, head."

This celestial image of the Triumphant Warrior appears (real-time) in the backdrop of the sky, confirming the devil's defeat. These magnificent events—the crucifixion and the resurrection scenes—are astronomically connected; one constellation portrays Yeshua, as the Lamb of God, the other—the Warrior/Gibbôr!

The enemy's head was crushed at the crucifixion of Messiah. During Messiah's resurrection, the crushed head of the enemy is symbolically carried away (as portrayed by the constellation Perseus). These events fulfill both the Gen 3:15 and 1 Sam 17:54 prophecies regarding the Messiah's total triumph over the enemy of our souls. These constellations—depicting the crucifixion and resurrection—have similar positions in the backdrop of the sky over Jerusalem during other proposed dates of the crucifixion/resurrection as AD 27 through AD 34.

Evening and morning hours are especially significant in the Bible, as are the positions of the constellations at these times. "And the evening [darkness] and the morning [light] were the first day" (Gen 1:5). The Scriptures say: "Let my prayer be set before You as incense, the lifting up of my hands as the evening sacrifice" (Ps 141:2). "O God, You are my God; early will I seek You" (Ps 63:1). "My soul waits for the Lord more than those who watch for the morning—yes, more than those who watch for the morning" (Ps 130:6).

"Through the Lord's mercies we are not consumed, because His compassions fail not. They are new every morning; great is Your faithfulness" (Lam 3:22, 23).

The celestial sky displays God's cosmic clock which enables us to know and keep the Lord's appointed feast days. God's cosmic clock is precise to the second.

"Then God said, 'Let there be lights in the firmament of the heavens to divide the day from the night; and let them be for signs and seasons, and for days and years'" (Gen 1:14).

According to the first chapter of Genesis, one of the primary purposes in observing the starry universe is to understand its signs and seasons. The Hebrew word sign is אוֹת/*oth*, meaning "a signal, a beacon." The Hebrew word for seasons is מוֹעֵד/*moed*, or defined as "an appointment, a fixed time, a festival."

The word used here for seasons, מוֹעֵד/*moed*, does not mean the four seasons, but rather the Lord's feast days. The Lord's feast days are the weekly Shabbat (Sabbath), and the annual feast days: Passover, Unleavened Bread, Firstfruits, Pentecost, the Feast of Trumpets, Day of Atonement, and the Feast of Tabernacles.

Yeshua instructs us to pay attention to the signs of the heavens. In doing so, we can watch for the cosmic signs of his second coming and point others to the gospel. "And there will be signs in the sun, in the moon, and in the stars" (Luke 21:25).

As we turn the cosmic clock back to the crucifixion and resurrection of the Messiah, we are viewing the Creator's אוֹת/*oth*, "a signal, a beacon" on the Lord's מוֹעֵד/*moed*, or "feast days," specifically Passover, Unleavened Bread, and Firstfruits of AD 31. Each year the constellations celebrate the spectacular gospel story. The Almighty is declaring his prophetic, pictorial poetry scripted in heaven—and it is being fulfilled on Earth!

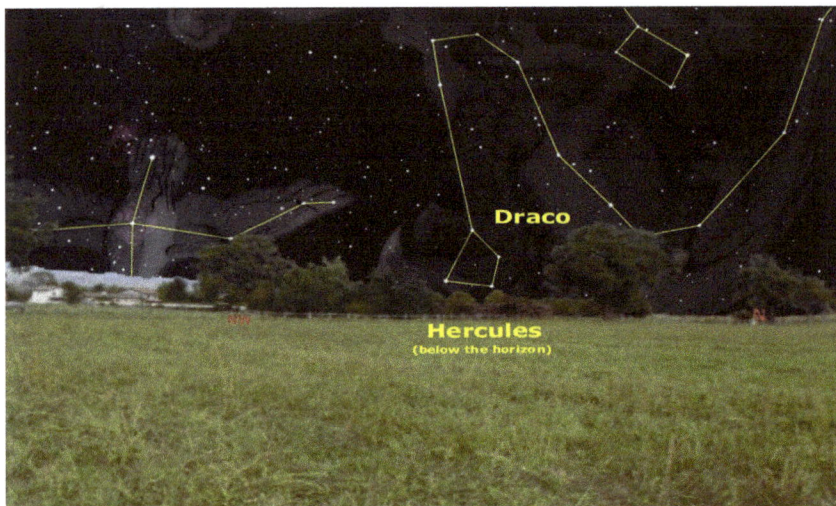

Stellarium Image (close-up): This astronomical configuration occurred April 25, AD 31, during Passover at the crucifixion of Messiah. At 12:00 noon the head of Draco began descending below the northwest horizon, disappearing between 2:00 to 3:00 p.m. in the afternoon. Draco's head represents Satan's authority being crushed beneath the feet of the Great Gibbôr, Messiah Yeshua. Hercules is located beneath the northwest horizon (symbolically) accomplishing his feat of strength!

Yeshua Gibbôr, the "Dragon slayer" rescues us from all serpentine powers.

A scripture reference for this concept of conquest is found in the book of Hebrews: "That through death He might destroy him who had the power of death, that is, the devil" (Heb 2:14).

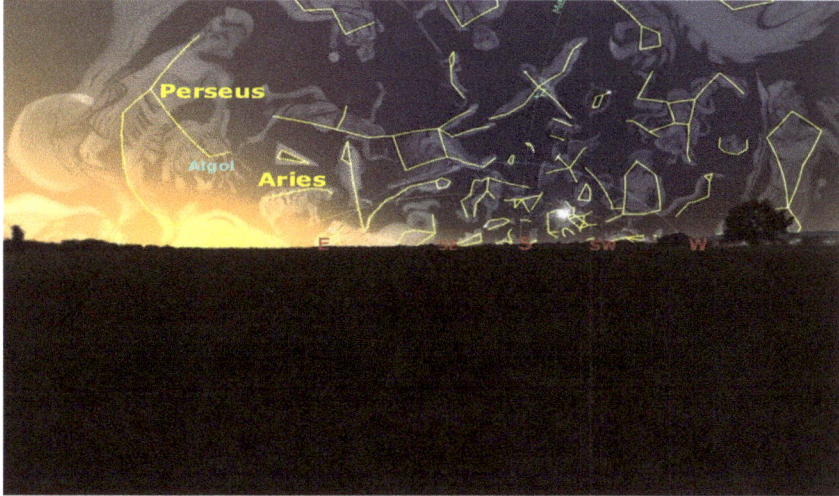

Stellarium image: This astronomical configuration occurred on Sunday morning, April 29, AD 31, at 4:45 a.m. The constellations Aries and Perseus rise at 4:30 a.m. just before sunrise, depicting Satan's head (Al Gol) in the left hand of Perseus. Perseus in Hebrew means "breaker" and the star *Al Gol* is an apparent abbreviation of (Gulgoleth) the Hebrew word for (skull, head.) This imagery indicates the headship authority of Satan has been utterly broken and removed from the people of God. There are several other significant signs occurring here. Yeshua has indeed conquered Satan and death!

The course of the constellations is calibrated by the Creator to serve as a witness of the death, burial, and resurrection of Yeshua. The heavens are, indeed, proclaiming the everlasting gospel! (Ps 19:1). When we look up and observe the Messiah crushing the head of the serpent/dragon may we grasp this glorious sign and God's signature in the stars.

My star song tribute, "The Mighty One," magnifies the message of the constellation Hercules.

The Mighty One

(Isa 9:6; Ps 91:13)
By Bruce J. Patterson

Chorus
The Mighty One crushed the enemy
The Mighty One won the victory—repeat

Verse 1
He is the Mighty God
The great King of Zion
He is the Mighty God
Great King of Zion

Chorus

Verse 2
Every foe will be brought low
By our Mighty Hero
Every foe will be brought low
By our Hero

Chorus

Verse 3
In the garden he knelt down
His blood fell to the ground

In the garden he knelt down
His blood fell to the ground

Chorus

Verse 4
On the cross he was bruised
He thrust to crush the dragon's head
On the cross he was bruised
He crushed the dragon's head

Chorus

Constellation Name	Translation
Gibbôr (Hebrew) גִּבּוֹר	strong, mighty, hero, champion
Hercules (Latin)	inferred: strength
Engonasī (Latin)	the kneeler, the constellation Hercules
Engonasin (Greek)	the kneeler
wǔ xiān zuò (Chinese)	the immortal martial constellation

Enclosed are derivatives and similar words in ancient languages which help clarify the core meaning of this constellation/star name.

Ancient Hebrew/ Aramaic	Phonetic Spelling	Summarized Meaning
Hebrew חַיִל	*chayil*	(strength, might, efficiency, wealth, army) From *chul* or *chil*: to be firm, strong

Ancient Gentile Language	Phonetic Spelling	Summarized Meaning
Greek	*hḗrōs*	hero
Greek	*kléos*	renown, glory

Ancient Gentile Language	Phonetic Spelling	Summarized Meaning
Arabic	*ḥayl*	strength
Assyrian	*khé la*	strength, power, force
Tigrinya/Ethiopian	*ḣayli*	strength
Arabic	*Al-Jathi*	the kneeling (man)

Scripture References

Gen 3:15; John 3:16, 8:8; Gen 2:7; Jer 31:33; Heb 10:16; Ps 113:6; John 13; Eph 5:25–27; Luke 22:41–44, 13:16; Acts 2:24; Ps 16:9–10; John 12:31, 14:30, 16:11; Acts 3:17, 4:8, 4:26

The Stars

Ras Al Gethi

THE STAR NAME RAS Al Gethi (pronounced ras-al-GAYTH-ee) is Arabic for "head of the kneeler." Scripture declares Yeshua as the *head* and Savior of the body—the church, the spiritual Israel of God, Bride of the Messiah. Scripture also teaches that Yeshua is the head of all principality and power; his headship authority is above every power that exists—visible and invisible.

Gibbôr, the Kneeler, represents the greatest act of strength of all time, an exposition of the Messiah's accomplishments on behalf of his people.

The following lessons from the book of Colossians, First John, Hebrews, and Isaiah help us understand the authority of Messiah:

> And He is the head of the body, the church, who is the beginning, the firstborn from the dead, that in all things He may have the preeminence. (Col 1:18)

> And you are complete in Him, who is the head of all principality and power. (Col 2:10)

> Having wiped out the handwriting of requirements that was against us, which was contrary to us. And He has taken it out of the way, having nailed it to the cross. Having disarmed principalities and powers, He made a public spectacle of them, triumphing over them in it. (Col 2:14, 15)

The word "handwriting," or *cheirographon* in the Greek vernacular, is defined as "a (hand written) legal document or bond indicating that men must be charged for the offences they commit, and pay the penalty." *Cheirographon* was positioned in the center of the court or assembly by the accusing witness. Our *cheirographon* or "record of sin" was nailed to the cross with the Messiah. So fully does God forgive us, he not only removed our sins, he removes even the "record" of our sins.

"Having disarmed principalities and powers . . ." The Greek word "disarmed" is *apekduomai*: "to strip off from oneself; to completely strip off." Yeshua (completely) stripped the enemy of his power and authority—for those who receive him.

"He who sins is of the devil, for the devil has sinned from the beginning. For this purpose, the Son of God was manifested, that He might *destroy* the works of the devil" (1 John 3:8).

Luó (λύω), the Greek word for "destroy," means "to loose, to release, to dissolve." Those who receive Yeshua are immediately released from the works of the enemy—the devil's hold on our lives. As disciples of Yeshua, we are given this same mandate. And in Yeshua's name, we are empowered to destroy the works of the devil in our lives and in the lives of others.

"Inasmuch then as the children have partaken of flesh and blood, He Himself likewise shared in the same, that through death He might destroy him who had the power of death, that is, the devil" (Heb 2:14).

This Greek word for destroy, *katargeó*, means "to render inoperative." "To abolish" means "to make completely inoperative or, to put out of use." The devil's works are rendered inoperative in the lives of those who receive Yeshua. Satan is a defeated foe. The enemy only can gain a foothold if we give him place.

Isaiah says: "In that day the Lord with His severe sword, great and strong, will punish Leviathan the fleeing serpent, Leviathan that twisted serpent; and He will slay the reptile that is in the sea" (Isa 27:1).

The Hebrew word for "punish" (Isa 27:1) is *paqad*, "to attend to, visit, muster, appoint." The Lord "attended to" the enemy by means of the cross when he crushed the devil's head of authority. When he returns, he will complete his task by destroying the enemy's very existence.

In Isa 27:1, *Leviathan, that twisted serpent* (עֲקַלָּתוֹן נָחָשׁ לִוְיָתָן), corresponds to the constellation *Draco*. The Hebrew word for *Leviathan*, לִוְיָתָן/ *Livyathan*, means "serpent; a sea monster or dragon." The Hebrew word for twisted עֲקַלָּתוֹן *aqallathon*, means "crooked." The Hebrew word for "serpent" is נָחָשׁ/*nachash*, which means "a serpent." Gibbôr, the Kneeler, is depicted as crushing Draco, the serpent, rendering him inoperative.

In Isa 27:1, (1) *Leviathan, the fleeing serpent* corresponds to the constellation *Hydra*. (2) *Leviathan, that twisted serpent* corresponds to the constellation Draco. (3) The *reptile in the sea* corresponds to the constellation *Cetus*. Each of these three constellations convey different aspects of the enemy.

Our great *Gibbôr*, Yeshua, has conquered death, hell, and the grave. He destroyed the curse of the enemy over our lives including sin, sickness, disease, poverty, fear, apprehension of the future, demonic activity (Deut 28; Gal 3:13). He destroyed the headship authority of the enemy from the lives of those who receive the gift of salvation and learn to live as his disciples (Col 1:13–14, 2:15; Heb 2:14–15).

The apostle Paul offers insight as to how we are to practically implement the Messiah's victory in our lives: "For your obedience has become known to all. Therefore, I am glad on your behalf; but I want you to be wise in what is good, and simple concerning evil. And the God of peace will crush Satan under your feet shortly" (Rom 16:19–20).

One of the believers' greatest weapons is peace (shalom). The devil cannot be successful against us when we walk in God's (shalom) peace.

God calls us to be wise in what is good and simple, or "innocent, blameless," concerning evil, or *kakía*, defined as "inner malice." The Greek word "shortly" is *tachos*, or "quickness, speed, hastily, immediately."

In Messiah, we have peace with God—and the peace of God which passes all understanding. The Hebrew word for peace, *shalom*, is defined as "nothing missing, nothing lacking; having complete well-being (in the Lord)." And God provides us with spiritual armor: "having shod your feet with the preparation of the gospel of peace" (Eph 6:15).

"And the God of peace will crush Satan under your feet shortly" (Rom 16:19–20). Implementing of the word of God in our lives puts the enemy under our feet.

The enemy does not belong in our bodies, marriages, homes, businesses, or wallets. The enemy belongs beneath our feet. With God's help we can enforce this reality.

God is accelerating his kingdom authority in our day. Remember, there are more with us than against us. "Do not fear, for those who are with us are more than those who are with them" (2 Kgs 6:16).

> Finally, my brethren, be strong in the Lord and in the power of His might. Put on the whole armor of God, that you may be able to stand against the wiles of the devil. For we do not wrestle against flesh and blood, but against principalities, against powers, against the rulers of the darkness of this age, against spiritual

hosts of wickedness in the heavenly places. Therefore, take up the whole armor of God, that you may be able to withstand in the evil day, and having done all, to stand. Stand therefore, having girded your waist with truth, having put on the breastplate of righteousness, and having shod your feet with the preparation of the gospel of peace; above all, taking the shield of faith with which you will be able to quench all the fiery darts of the wicked one. And take the helmet of salvation, and the sword of the Spirit, which is the word of God; praying always with all prayer and supplication in the Spirit, being watchful to this end with all perseverance and supplication for all the saints. (Eph 6:10–18)

When we follow the biblical example of ancient Israel's command by God to inherit the promised land, we discover the exceeding great and precious promises God has made available to us in the Messiah today.

The God of Israel reserved the land of Israel for the descendants of Abraham, Isaac, and Jacob. This covenantal promise is good for a thousand generations (Ps 105:8–11). Jewish believers in the Messiah can experience both the physical and spiritual promised land. Israel, the physical promised land, is emblematic of the spiritual promised land. God's spiritual land—his exceeding great and precious promises—is a reality today. This habitation is the "land" where all the promises of God are yes and amen, to the glory of God in Yeshua, the Messiah (2 Cor 1:20; 2 Pet 1:4). The Bible tells us God brought Israel out of Egypt that he might bring them into the promised land (Deut 6:23).

And so it is with us today; God separates us unto himself so he can conform us to the image of the Messiah, thereby enabling us to inherit his promises in this life and in the world to come. As we follow in the steps of the Messiah, we can realize we are well able to possess this good land.

Israel's possession of the physical promised land was a type of the spiritual promised land we can enter today: "Every place on which the sole of your foot treads shall be yours" (Deut 11:24). Each step we take can become an act of worship to God—and a crushing blow to the enemy beneath our feet.

The Almighty desires that we become strong in him and in the power of his might. He desires that we become little *gibbôrs* in the Messiah, our great Gibbôr.

God enables his people to be strong and to carry out great exploits, because "greater is He who is in you than he who is in the world" (1 John 4:4; John 14:12).

"Those who do wickedly against the covenant he shall corrupt with flattery; but the people who know their God shall be strong, and carry out great exploits" (Dan 11:32).

The heavens are declaring the glory of God! Yᵉhôvâh's Messiah has triumphed over all the power of the enemy both in heaven and Earth.

Halleluyah to the Messiah, Yeshua, our Mighty Hero!

Conventional Star Name/Meaning	Name/Meaning Confirmed or Corrected
Ras Al Gethi—the head of him who bruises	confirmed: head of the kneeler

Enclosed are derivatives and similar words in ancient languages which help clarify the core meaning of this constellation/star name.

Ancient Hebrew/ Aramaic	Phonetic Spelling	Summarized Meanings
Hebrew ראשׁ	*rosh*	head

Ancient Gentile Language	Phonetic Spelling	Summarized Meanings
Arabic	*Al-Jathi*	the kneeling (man)
Arabic	*ra's*	head
Arabic	*jaṯā* (phonetic—*getha*)	to kneel
Persian	*jāyi*	kneel
Persian	*jāṣī*	kneeling
Hindi	*ghutne*	kneel
Sanskrit	*ganti*	kneel
Chinese	*guì*	kneel
Egyptian	*sai*	bow down

Scripture References

Col 1:18, 2:10–15; 1 John 3:8; Heb 2:14; Isa 27:1; Rom 16:19–20; Ps 19; Eph 6:10–18; 2 Pet 1:4; 2 Cor 1:20; Deut 11:24; Dan 11:32

Kornephoros

The star name Kornephoros (pronounced core-ne-FOR-uss) is a combination of two Greek words, *korunē*, meaning "club, mace, shepherds' staff," and *phoréas* or *foréas*, meaning "bearer." When combined, these words mean "club-bearer."

One of the Hebrew words for club is שֵׁבֶט/*shebet*, which means "rod, staff, club, scepter, tribe." This analogy agrees with the biblical narrative of the Messiah as Shepherd and King. The *shebet*, symbolically, is held by the Messiah in both offices.

A prime example of the word *shebet* is found in the classic psalm: "Your rod [שֵׁבֶט/*shebet*] and Your staff [מִשְׁעֶנָה/*mish-ay-naw*], they comfort me" (Ps 23:4).

It is also used in the Proverbs: "He who spares his rod [שֵׁבֶט/*shebet*] hates his son, but he who loves him disciplines him promptly" (Prov 13:24).

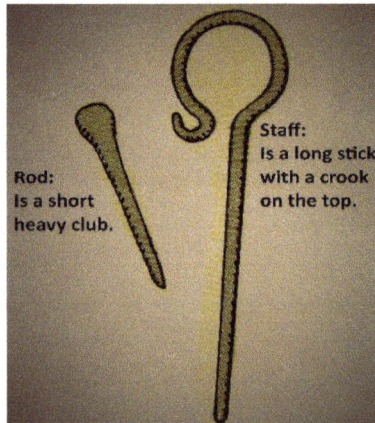

A rod (שֵׁבֶט/*shebet*) is used for correction, discipline, and guidance while a staff (הַמִּשְׁעֶנָה/*mish-ay-naw*) was used as a walking staff and defensive weapon. (Drawing by Jim Pinkoski.)

The *shebet* is a shepherd's instrument used to guide, direct, correct, and if necessary, strike his sheep. The *shebet* was also used to lovingly discipline children. This agrees with Prov 3:12, as quoted in the book of Hebrews: "For whom the Lord loves He chastens, and scourges every son whom He receives" (Heb 12:6 and Prov 3:11–12).

The Hebrew Bible uses imagery to depict God as examining his people describing them as "sheep." The examination, "to pass under the rod," is a figure of speech describing a shepherd's way of counting and examining his flock (Lev 27:32; Jer 33:13; Mic 7:14). Spiritually speaking, it is by this symbolic examination the people were brought into the promised land. They were selected and reconstituted as God's own covenant people.

The rod (שֵׁבֶט/*shebet*) was, in effect, a two-foot club functioning as a tool under which sheep passed. The shepherd would count the sheep in his flock, making sure they were present and accounted for. The rod (שֵׁבֶט/*shebet*) also symbolizes the process of "undergoing close scrutiny" by the word of God. A shepherd would run his rod across the sheep in order to inspect them for ticks and injuries beneath the wool on the sheep's skin. God is illustrating, by means of his rod (שֵׁבֶט/*shebet*), that he is giving his people careful scrutiny in order to count and evaluate their quality (Heb 4:12, 13).

According to artist renderings of the zodiac, Hercules is depicted as holding a club in his right hand. The star name Kornephoros (club-bearer) may also signify a rod of iron (*shebet* of iron) as depicted in the Second Psalm—a description of the Messiah's triumph and kingdom. "Ask of Me, and I will give You The nations for Your inheritance, and the ends of the earth for Your possession. You shall break them with a rod [שֵׁבֶט/*shebet*] of iron; You shall dash them to pieces like a potter's vessel" (Ps 2:8, 9).

The functionality of the "*shebet* of iron" is further expanded by the prophet Daniel: "And in the days of these kings the God of heaven will set up a kingdom which shall never be destroyed; and the kingdom shall not be left to other people; it shall break in pieces and consume all these kingdoms, and it shall stand forever" (Dan 2:44). Here, we understand the kingdoms of this world will eventually come to ruin but the kingdom of the Messiah continues into the millennial reign and beyond. Where will you stand?

Now is your opportunity to make the decision to choose Yeshua/Gibbôr as your Lord and Savior. If you choose him now, it's because he has already chosen you. Allow Yeshua to become your Shepherd, Messiah, and King. He knows how to take care of you and provide for your needs more completely than you could ever realize alone.

Yeshua (spiritually) raises us from the dead and will one day raise us to immortality. We, who are alive until his coming, will put on immortality—in the twinkling of an eye.

A beautiful picture of the love and power of the living Shepherd is shown in the Gospel of Mark when Yeshua raises the daughter of the ruler of the synagogue from the dead: "Then He took the child by the hand, and said to her, 'Talitha, cumi,' translated, 'Little girl, I say to you, arise.' Immediately the girl arose and walked, for she was twelve years of age. And they were overcome with great amazement" (Mark 5:41, 42).

The Aramaic name *Talitha* means "a lamb" and *cumi* means "arise."

Let Yeshua take you by the hand and you shall live!

Yeshua Gibbôr must reign with rod (שֵׁבֶט/*shebet*) in hand until the last enemy (death) dies beneath his kingship authority. Yeshua Gibbôr will gently guide his sheep until the last little lamb passes safely under the rod (שֵׁבֶט/*shebet*) of his protection into the fold of the eternal redeemed.

Conventional Star Name/Meaning	Name/Meaning Confirmed or Corrected
Kornephoros—the branch kneeling	corrected: club bearer

Enclosed are derivatives and similar words in ancient languages which help clarify the core meaning of this constellation/star name.

Ancient Hebrew/ Aramaic	Phonetic Spelling	Summarized Meaning
Hebrew שֵׁבֶט	*shebet*	rod, staff, club, scepter, tribe

Ancient Gentile Language	Phonetic Spelling	Summarized Meaning
Latin	*phorus*	bearing
Latin	*corona*	crown

Ancient Gentile Language	Phonetic Spelling	Summarized Meaning
Tamil	korōṉā	crown
Greek φορέας	phoréas or foréas	bearer
Greek κορώνα	koróna	crown, koruna, krone
Greek κορυνηφόρος	korunēphóros	club or staff bearer
Ancient Greek	korunē	suffixed form *koru-nā. from Greek korunē, club, mace a shepherd's staff
German and cognate	krone	from Czech and Slovak koruna; cognate with krone and German Krone
Greek ρόπαλο	rópalo or apelatíki	bat, club, mace, weapon

Scripture References

Gen 49:10; Ps 23:4; Prov 13:24; Heb 12:6; Prov 3:11–12; Heb 4:12–13; Lev 27:32; Jer 33:13; Mic 7:14; Ps 2:8–9; Dan 2:44, Ps 110:1

13

The Constellation: Sagittarius

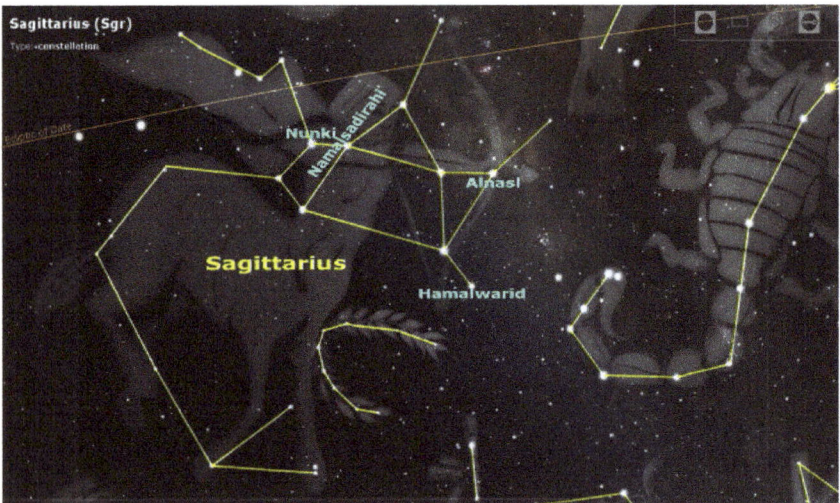

THE CONSTELLATION SAGITTARIUS REPRESENTS the "God Man" who descended from heaven and became a man in order to restore us to the Father of life.

The word *sagittarius*, defined as "archer" in Latin, can be seen in the southern celestial heavens. *Qesheth*, its Hebrew name, is also defined as "archer." Additionally, the Sumerian name for Sagittarius, "Nunki," is a combination of two words: *nun*, meaning "prince," and *ki*, meaning "earth," or "prince of the earth."

Sagittarius is a centaur whose "horse and man" characteristics symbolize those of the Messiah. The rider on this horse represents the immortal Son of Man, the Messiah. The horse represents Messiah's administration, carrying his message to the world.

Sagittarius, the Centaur, represents the dual nature of the Messiah as God and man. The Bible for example reveals the *dual nature* of the Messiah in the ark of the covenant made of acacia wood overlaid with pure gold. The acacia wood represents the humanity of the Messiah while its pure gold overlay represents his deity.

Sagittarius is the story of the Triumphant Archer going forth conquering, as depicted in the book of Revelation: "And I looked, and behold, a white horse. He who sat on it had a bow; and a crown was given to him, and he went out conquering, and to conquer" (Rev 6:2).

A key scripture corresponding to Sagittarius (*Qesheth*) is found in Ps 45:

> I recite my composition concerning the King;
> My tongue is the pen of a ready writer.
> You are fairer than the sons of men;
> Grace is poured upon Your lips;
> Therefore, God has blessed You forever.
> Gird Your sword upon Your thigh, O Mighty One,
> With Your glory and Your majesty.
> And in Your majesty ride prosperously because of truth, humility, and righteousness;
> And Your right hand shall teach You awesome things.
> Your arrows are sharp in the heart of the King's enemies;
> The peoples fall under You.
> Your throne, O God, is forever and ever;
> A scepter of righteousness is the scepter of Your kingdom.
> You love righteousness and hate wickedness;
> Therefore God, Your God, has anointed You
> With the oil of gladness more than Your companions. (Ps 45:1–7)

Sagittarius depicts the archer shooting his arrows at the heart of the enemy, Scorpio, the Scorpion. Antares, the brightest star in Scorpio, is defined as anti-Aries, or, against Aries (the Ram). The name Antares corresponds, biblically, to the Greek term "antichrist" or, in Hebrew, "anti-Messiah," defined as "an opponent of the Messiah."

In Sagittarius, we observe the prophetic inferences of the Messiah and his kingdom administration destroying the lying works of the antichrist spirit by aiming and shooting his sharp arrows of truth straight into the heart of this foe. Arrows, in the Bible, represent the descendants of the righteous brought up in the nurture and admonition of the Lord: "Like arrows in the hand of a warrior, so are the children of one's youth. Happy is the man who has his quiver full of them; they shall not be ashamed but shall speak with their enemies in the gate" (Ps 127:4, 5).

The constellation Sagittarius depicts the administration of the Messiah's triumph over evil. This victory can be celebrated by all who participate in the ancient cosmic war. "But God shall shoot at them with an arrow; suddenly they shall be wounded. So He will make them stumble over their own tongue; all who see them shall flee away. All men shall fear, and shall declare the work of God; for they shall wisely consider His doing. The righteous shall be glad in the Lord, and trust in Him. And all the upright in heart shall glory" (Ps 64:7–10).

Epic Discovery

A full lunar eclipse/blood moon (in the Scales/claws) occurred on Passover, April 16, 1448 BC, at midnight. This Passover blood moon may have occurred during the exodus or one year after the exodus, when Israel was at Mount Sinai in Arabia.

On April 25, AD 31, the night following the crucifixion of the Messiah, the Moon virtually rose (at 7:00 p.m.) in the same position in Libra as it did on Passover, April 16, 1448 BC. Could the Moon confirm the date of the Passover crucifixion? It very well may!

I support the hypothesis which—through scientific research methods—concludes the Messiah was crucified on Wednesday, April 25, AD 31 (Nisan 14), and resurrected three days later—after sundown on April 28, AD 31. When the women arrived at the tomb—early Sunday morning (on the Feast of Firstfruits, April 29)—the Messiah had already risen!

A blood moon occurs when the Sun, Earth, and Moon align and the Moon passes into Earth's shadow, casting a reddish glow on the Moon. A partial lunar eclipse can cause the Moon to cast a reddish hue.

The prophet Joel prophesied regarding blood moons and solar eclipses; hence, the apostle Peter quoted this verse on Shavuot/Pentecost fifty days after resurrection of Yeshua: "I will show wonders in heaven above and signs in the earth beneath: Blood and fire and vapor of smoke. The sun shall be turned into darkness, and the moon into blood, before the coming of the great and awesome day of the Lord" (Acts 2:19–20; Joel 2:28–32).

Joel's prophecy, quoted by the apostle Peter on Shavuot/Pentecost, indicates a blood moon was seen within the context of the Passover crucifixion. This blood moon lunar eclipse occurred April 25, AD 31. Peter's testimony helps identify AD 31 as the year of the Messiah's crucifixion. Note: According to Stellarium, Passover lunar eclipses (blood moons) of AD 32 and AD 33 were not visible from Jerusalem. The only visible Passover lunar eclipse that occurred from AD 27 to AD 34 is the AD 31 partial blood moon, as seen from Jerusalem (see "Chart of Passover Crucifixion Dates," Appendix 1)

"Moonrise" occurs about fifty minutes later each day. In biblical astronomy, the Moon represents the church, "the spiritual Israel of God." The Moon has no light of its own but reflects the light of the Sun. Likewise, we are to reflect the light of Yeshua (the Sun of Righteousness) in a dark world. Blood moons may represent the blood of the martyrs. Notably, Passover blood moons represent the atoning blood of the Messiah covering and redeeming his people. A blood moon, in the Scales and claws of the Scorpion, represents the Creator, Redeemer snatching his people from the claws of the enemy. The Moon occasionally traverses the scales of Libra on Passover. However, a (full or partial) Passover blood moon occurring in Libra is rare.

The Moon is called "the faithful witness in the sky" (Ps 89:37). In AD 31, the biblical new year began on Nisan 1 (April 11). On this day, the Sun, new Moon, Jupiter, and Mercury were in the context of Aries, the Ram, collectively conveying their messages according to biblical astronomy:

1. The Sun represents Yeshua, the light of the world (highlighting its theme).

2. The Moon represents God's people reflecting the light of the Son.

3. The new Moon represents new birth.

4. Jupiter in Hebrew is צדק/tzedek, meaning "righteousness."

5. Mercury denotes "swift messenger."

My interpretation may be stated as: "God gives his people new birth through the righteous Lamb of God."

Could the Moon's position and path reveal the actual three days and three nights (seventy-two hours) the Messiah was in the tomb? Wednesday, April 25, AD 31, at 7:00 p.m. approximately when the Messiah's tomb was sealed, a full Passover moon rose over Jerusalem after sundown. Two and a half hours later a partial lunar eclipse of the Moon appeared over Jerusalem. This lunar eclipse (blood moon) peaked at 10:31 p.m. The duration of the eclipse lasted 115 minutes: 9:35 to 11:30 p.m. (1 hour and 55 minutes). This partial blood moon could be symbolic of the first-century redeemed remnant of Israel. About a third of the Moon was totally immersed in the Earth's shadow, casting a reddish hue. Atmospheric dust may have intensified its redishness.

When the Messiah was crucified, the Moon was positioned in Libra, the Scales. Libra represents righteousness and justice: "Righteousness and justice are the foundation of Your throne; mercy and truth go before Your face" (Ps 89:14). The Moon moved out of Libra and "passed over" (illustrating Passover) Scorpio, the Scorpion (representing Satan). Three days later the Moon moved into Sagittarius, the Archer.

One of the major themes of Sagittarius is the offspring of the redeemed. The Moon traveled through these three constellations while the Messiah's body laid in the tomb—thus tracking (seventy-two hours) three days and three nights. It appears God used the Moon (the faithful witness in the sky) to providentially declare the greatest story ever told in the following three stages:

1. God confirmed the Passover crucifixion (April 25) with the visible partial blood moon in Libra.

2. God marked the first day of Unleavened Bread (Nisan 15) *a Holy Convocation/High Day-Sabbath* with the Moon "passing over" (illustrating Passover) Scorpio, the Scorpion.

3. God marked Firstfruits (April 29, "resurrection day") with the Moon in Sagittarius, placing an emphasis on its meaningful context. In biblical typology the number three denotes resurrection.

Two-star names in Sagittarius (the Archer) enhance the theme of Messiah's spiritual offspring: Nun in Nunki, meaning "sprouting seed, offspring," and Al Nasl, meaning "offspring, or, descendants."

Do the heavens declare that Yeshua has redeemed his people as those who endure as "warriors" of God's justice in the earth?

Hear the word of the Lord: "My covenant I will not break, nor alter the word that has gone out of My lips. Once I have sworn by My holiness; I will not lie to David: His seed shall endure forever, and his throne as the sun before Me; it shall be established forever like the moon, even like the faithful witness in the sky. Selah" (Ps 89:34–37).

The following poignant scripture has to do with the Messiah's suffering in order to redeem us—his spiritual offspring: "Yet it pleased the Lord to bruise Him; He has put Him to grief. When You make His soul an offering for sin, He shall see His seed, He shall prolong His days, and the pleasure of the Lord shall prosper in His hand. He shall see the labor of His soul, and be satisfied. By His knowledge My righteous Servant shall justify many, for He shall bear their iniquities" (Isa 53:10–11).

The signs in the Sun, Moon, and stars collectively identify the time frame of the death, burial, and resurrection of the Messiah.

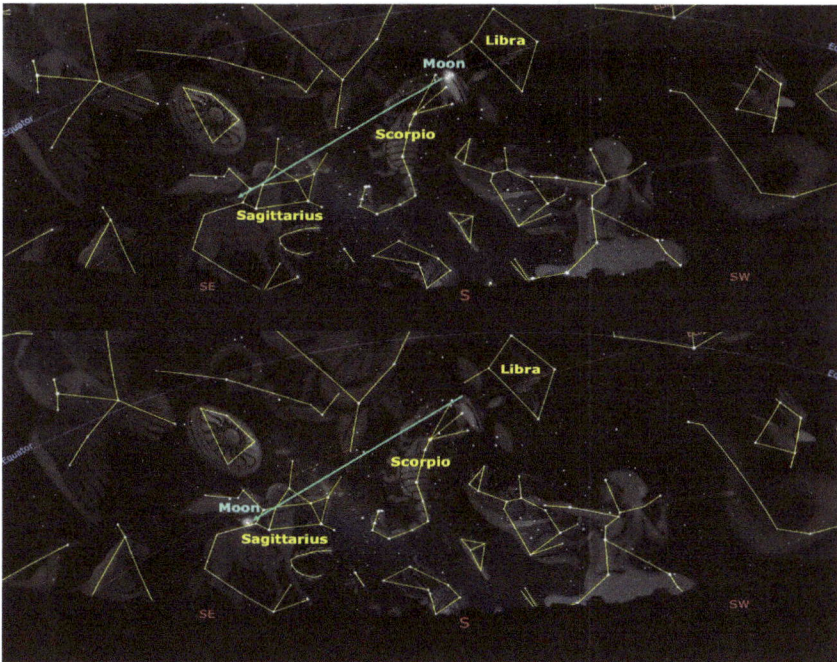

TOP: Image A. BOTTOM: Image B.
Stellarium image: image A configuration occurred April 25, AD 31, following the crucifixion Passover of the Messiah when the Moon was in Libra, the Scales. The Moon moved out of Libra and "passed over" (illustrating Passover) Scorpio, the Scorpion (representing Satan), and three days later moved into Sagittarius, the Archer (image B).

According to E. W. Bullinger, the Greek poet Aratus, known for his poem on the subject of Sagittarius, writes, "Midst golden stars he stands refulgent now, and thrusts the scorpion with his bended bow."[1]

Thus, my classical guitar star song for Sagittarius, "The Triumphant Archer," summarizes this story of the gospel.

The Triumphant Archer

(Ps 45:3–5)
By Bruce J. Patterson

The constellation of Sagittarius (the Archer). (Image by Amy Manson, 1895; from *The Gospel in the Stars*, by Joseph A. Seiss.)

Chorus
The Triumphant Archer, riding prosperously
With his bow and arrows, aiming and speeding
At the heart of the foe
The Triumphant Archer, standing for the defense
And administration of God's righteousness
The delightful, gracious, Triumphant Archer

Verse 1
A mounted warrior riding as a king
Armed with bow and arrows, shooting down his enemies
The Mighty One girding himself with honor and majesty
The conqueror going forth conquering

---

1. Aratus, quoted in Bullinger, *Witness of the Stars*, 54.

Chorus

Verse 2
Once humbled on the cross, now he's exalted
He tasted death for us all so we might live
Midst the golden stars he stands, refulgent now
And thrusts the foe with his bended bow

Chorus

Constellation Name	Translation
Sagittarius (Latin)	an archer, bowman
Qasat (Hebrew)	archer
Toxeutês (Greek)	the archer
Ar-Rami (Arabic)	the archer
Nunki (Sumerian)	prince of the earth
rén mǎ zuò (Chinese)	the horse-man constellation

Enclosed are derivatives and similar words in ancient languages which help clarify
the core meaning of this constellation/star name.

Ancient Hebrew/ Aramaic	Phonetic Spelling	Summarized Meaning
Hebrew קֶשֶׁת	qaṣắṭ	archer (Hebrew name of archer constellation)
Hebrew קֶשֶׁת	qesheth	archer
Hebrew חֵץ	khayts	arrow

Ancient Gentile Language	Phonetic Spelling	Summarized Meaning
Latin	Sagittārius	an archer, bowman
Assyrian	qash ' šat ṭa ya	archer
Arabic	al Kaus	bow

Ancient Gentile Language	Phonetic Spelling	Summarized Meaning
Sumerian	*Nunki*	prince of the earth
Sumerian	*Nun*	prince
Sumerian	*ki*	earth

Scripture References

Gen 21:20, 49:24; Pss 45:1–7, 64:7–10, 127:4–5; Hab 3:9; Rev 6:2, 19:11

The Stars

Naim (Namalsadirah)

THE STAR NAME NAMALSADIRAH is composed of the following three words: (1) The ancient Hebrew word נָעֵם/*naem*, meaning "pleasant, delightful, sweet"; (2) the Arabic definite article *al*, or "the"; (3) the word *sadirah*, likely from the Arabic word *ssadr*, meaning "chief, chairman." The combination of these three words implies "the delightful chief."

Some have associated the word *nama* with the Persian and Arabic word *na'ām*, or "ostrich." Although *na'ām* is phonetically similar to *nama*, it does not personify the narrative of Sagittarius, the Archer. The origin of the star name "Namalsadirah" is unknown. E. W. Bullinger does not include *sadirah* as part of this star name. He refers to the star as *Naim*.[1]

The Hebrew word נָעֵם/*naem* is mentioned in several Bible passages: "Behold, how good and how pleasant [נָעִים/*nâ'îym*] it is for brethren to dwell together in unity!" (Ps 133:1); "Praise the LORD! For it is good to sing praises to our God; for it is pleasant [נָעִים/*nâ'îym*], and praise is beautiful" (Ps 147:1).

When God's people dwell together in the unity of the Holy Spirit, there is true *naem*: pleasantness and delightfulness.

God is the author of all that is pleasant (*naem*), as revealed through this star name in the constellation Sagittarius.

---

1. Bullinger, *Witness of the Stars*, 53.

Conventional Star Name/Meaning	Name/Meaning Confirmed or Corrected
Naim—the gracious one	confirmed

Enclosed are derivatives and similar words in ancient languages which help clarify the core meaning of this constellation/star name.

Ancient Hebrew/ Aramaic	Phonetic Spelling	Summarized Meanings
Hebrew נָעֵם	naem	נְעִים/*nâ 'îym*, delightful (objective or subjective, literal or figurative):— pleasant, sweet. From נָעֵם/ *nâ 'êm*, naw-ame'; a primitive root; to be agreeable (literally or figuratively); pass in beauty, be delightful, be pleasant, be sweet
Hebrew נָעֵם	*nâ 'êm*	to be pleasant, be beautiful, be sweet, be delightful, be lovely Israelite woman: Naomi = "my delight"
Aramaic	*nyām*	portion
Hebrew סֵדֶר	*çeder*	to arrange; order
Aramaic	*sbr, sbrʾ (sāḇar, sāḇrā)*	master teacher
Aramaic	*sdr (sḏar (seḏer), seḏrā)*	to arrange
Aramaic	*sdrh*	manager

Ancient Gentile Language	Phonetic Spelling	Summarized Meanings
Phoenician נעם	*nm*	good
Arabic	*ni 'ma*	excellent, good
Arabic	*na 'ām*	ostrich
Egyptian Arabic	*naAaem*	ostrich
Persian	*na 'ām*	ostrich
Persian	*na-eem*	delight, pleasure
Persian	*nā 'īn*	pleasant, healthy

Ancient Gentile Language	Phonetic Spelling	Summarized Meanings
Persian	*na 'am*	being green and succulent (a branch) name of a woman, Naomi
Assyrian	*mahn'iana*	pleasant, pleasing
Arabic	*ṣāḥib*	chief, sovereign, king, master, leader
Arabic	*ssadr*	chief, chairman
Arabic	*ṣayid*	chief, governing leading
Persian	*ṣadr*	chief seat; prime minister; a judge, head, chief
Hindi	*ṣadr*	chief, chairman
Persian	*ṣadr*	returning
Akkadian	*sadāru*	to put in order, arrange in order
Assyrian	*si dra*	an array, a list, a row, a line, a string, an order, a series, an arrangement

Scripture References

2 Sam 23:1; Pss 16:6, 16:11, 27:4, 133:1, 147:1; Prov 22:18; Ruth 1:20; Song 1:16

Al Warida or (Hamalwarid)

The star name Hamalwarid combines two Arabic words: the Arabic word *hamal* means "male lamb" and Arabic word *warada* means "to arrive." Together, these words signify "arrival of the lamb," or "the lamb arrives." The word *hammāl* may be a derivative of the Arabic word *hamal* (male lamb), which is defined as "porter, bearer, to carry, carrier." This extended meaning, therefore, may imply, "the bearer (lamb) of (the message) will arrive."

The origin of the star name Hamalwarid is unknown. Bullinger does not include *Hamal* as part of this star name. It is simply known as Al Warida.

The Arabic word *hamal* may be a derivative of the ancient Hebrew word חָמַל/*châmal*, meaning "to spare, (have) compassion, (have) pity."

The second part of the word Hamalwarid (warid), is of the Semitic root *w-r-d* like the Hebrew יָרַד/*yarád*, "to go down, descend." When combined, these words may mean "(God) will come down and show compassion." A combination of Hebrew and Arabic implies: "the Lamb will arrive showing compassion."

The following scripture is an example of the use of the word חָמַל/*châmal* in the Hebrew Bible: "Then those who feared the Lord spoke to one another, and the Lord listened and heard them; so a book of remembrance was written before Him for those who fear the Lord and who meditate on His name. 'They shall be Mine,' says the Lord of hosts, 'On the day that I make them My jewels. And I will spare [חָמַל/*châmal*] them as a man spares [חָמַל/*châmal*] his own son who serves him'" (Mal 3:16, 17).

In the Bible, the Hebrew word יָרַד/*yârad* is used in reference to the tower of Babel: "But the Lord came down [יָרַד/*yârad*] to see the city and the tower which the sons of men had built. And the Lord said, 'Indeed the people are one and they all have one language, and this is what they begin to do; now nothing that they propose to do will be withheld from them. Come, let Us go down [יָרַד/*yârad*] and there confuse their language, that they may not understand one another's speech'" (Gen 11:5–7).

The Lord came down (יָרַד/*yârad*) on Mount Sinai: "Now Mount Sinai was completely in smoke, because the Lord descended [יָרַד/*yârad*] upon it in fire. Its smoke ascended like the smoke of a furnace, and the whole mountain quaked greatly" (Exod 19:18). The Greek Septuagint translation uses the word καταβαίνω/*katabainó* as "descended."

The following is an example of the word *katabainó*, defined as "to go down, or to descend" in the Greek translation of the book of John: "For I have come down [*katabainó*] from heaven, not to do My own will, but the will of Him who sent Me" (John 6:38). The Hebrew word יָרַד/*yârad* was most likely the word Yeshua chose to address the Jewish people.

"Religion" is man's attempt to reach God. True faith in God is realizing we cannot reach the God of heaven. Therefore, in his mercy, he came down (יָרַד/*yârad*) to Earth to reach us.

The Lord will come down (יָרַד/*yârad*) one more time when he returns for his saints:

> For the Lord Himself will descend from heaven with a shout, with the voice of an archangel, and with the trumpet of God. And the dead in Christ will rise first. Then we who are alive and remain shall be caught up together with them in the clouds to meet the Lord in the air. And, thus, we shall always be with the

Lord. Therefore comfort one another with these words. (1 Thess 4:16–18)

Conventional Star Name/Meaning	Name/Meaning Confirmed or Corrected
Al Warida—who comes forth	corrected: to descend and arrive inferred: (who comes forth)

Enclosed are derivatives and similar words in ancient languages which help clarify the core meaning of this constellation/star name.

Ancient Hebrew/ Aramaic	Phonetic Spelling	Summarized Meaning
Hebrew חָמַל	*châmal*	to spare: have compassion, (have) pity, spare.
Hebrew יָרַד	*yârad*	to go down, descend
Aramaic	*yrd*	to stream down

Ancient Gentile Language	Phonetic Spelling	Summarized Meaning
Arabic	*warada*	to arrive
Persian	*wārid*	one who comes, arrives, approaches; one who descends or enters, comers and goers a watering-place
Akkadian	*waradu*	descend
Phoenician	*yrd*	arrive
Semitic Root	*w-r-d* cognate to ירד.	יָרַד • (*yarád*) to go down, descend
Persian	*ḥamal*	a lamb; the sign Aries
Arabic	*ḥamal*	male lamb
Arabic	*ḥammāl*	porter, bearer, to carry, carrier

Ancient Gentile Language	Phonetic Spelling	Summarized Meaning
Persian	*ḥammāl*	a porter, carrier of burdens; a bearer
Turkish	*hamal*	bearer, carrier, porter
Egyptian Arabic	*hammael*	carrier
Assyrian	*ha mal*	porter, carrier, bearer
Hindi	*hhammal*	porter, bearer, to carry, carrier
Ethiopian Amharic	*hamal*	carrier

Scripture References

Gen 11:5–7, 28:12; Exod 19:18; Ps 72:6; Mal 3:16–17; 1 Thess 4:16–18

Nunki

The star name Nunki is a combination of two Sumerian words: *Nun*, meaning "prince" and *ki*, meaning "earth." Combined, these words mean "prince of the earth." The star name Nunki is also the Sumerian name for this constellation.

The Sumerian *nun* and the Hebrew *nun* may be derived from the same root. The ancient Hebrew *nun* is a picture of a sprouting seed, invoking the idea of continuing into a new generation, as in "to continue, perpetuation, offspring and heir."[2]

A classic example of the Hebrew meaning of *nun* is found in the personhood of Yeshua himself, who is called "the Son of David" and "the offspring of David": "I am the Root and the Offspring of David, the Bright and Morning Star" (Rev 22:16; Matt 21:9). There is only one to whom this title (Prince of the Earth) belongs, that is, the Lord Yeshua Hamashiach, the Messiah of Israel.

We must understand that the earth belongs to the Lord our Creator and Redeemer:

"The earth is the LORD's, and all its fullness, The world and those who dwell therein" (Ps 24:1).

---

2. Benner, *Ancient Hebrew Research Center-Nun*

"For thus says the LORD, Who created the heavens, Who is God, Who formed the earth and made it, Who has established it, Who did not create it in vain, Who formed it to be inhabited: "I am the LORD, and there is no other" (Isa 45:18).

As composed by William Chatterton Dix in his 1866 hymn titled "Hallelujah, Sing to Jesus," he *is* Earth's Redeemer: "Hallelujah! bread of heaven, here on earth our food, our stay: Hallelujah! here the sinful come to you from day to day. Intercessor, friend of sinners, earth's redeemer, plead for me, where the songs of all the sinless sweep across the crystal sea."[3]

Satan is called the prince of the world; he is not prince of the earth. "The earth is the LORD's, and the fullness thereof" (Ps 24:1). "The meek shall inherit the earth" (Matt 5:5).

This fallen world system is under the authority of Satan: "I will no longer talk much with you, for the ruler of this world is coming, and he has nothing in Me" (John 14:30). "The ruler of this world is judged" (John 16:11).

The Greek word for *world*, κόσμος/*kósmos* (kos'-mos), means "orderly arrangement, decoration; by implication, the world (in a wide or narrow sense), including its inhabitants, literally, figuratively, or, morally; adorning, world."

The book of First John defines the world system as such: "For all that is in the world—the lust of the flesh, the lust of the eyes, and the pride of life—is not of the Father but is of the world. And the world is passing away, and the lust of it; but he who does the will of God abides forever" (1 John 2:16,17).

All authority belongs to Yeshua, the true "Prince of Peace" (Isa 9:5). "And Jesus came and spoke to them, saying, "All authority has been given to Me in heaven and on earth. Go therefore and make disciples of all the nations, baptizing them in the name of the Father and of the Son and of the Holy Spirit, teaching them to observe all things that I have commanded you; and lo, I am with you always, even to the end of the age" (Matt 28:18–20).

Praise the Prince of Peace, praise the Prince of the Earth, Yeshua Hamashiach!

---

3. Dix, "Hallelujah, Sing to Jesus."

Conventional Star Name/Meaning	Name/Meaning Confirmed or Corrected
Nunki—Prince of the Earth	confirmed

Enclosed are derivatives and similar words in ancient languages which help clarify the core meaning of this constellation/star name.

Ancient Gentile Language	Phonetic Spelling	Summarized Meanings
Sumerian	*Nun*	prince
Sumerian	*ki*	earth

Scripture References

Ps 24:1; Matt 5:5; Isa 45:18; Rev 22:16; Matt 21:9; John 14:30, 16:11; 1 John 2:16–17; Matt 28:18–20

Al Nasl

The star name Al Nasl (pronounced all-NAH-zul) is a combination of the following two Arabic words: the Arabic definite article *al*, or "the," and the Arabic word *nasl*, meaning "offspring, or, descendants." The two words combined mean "the offspring." Nasl may also be a derivative of the Arabic *nuṣūl*, meaning "arrowhead." This narrative fits the constellation Sagittarius, the Archer.

Sagittarius depicts the Archer shooting his arrows at the heart of the enemy (Scorpio). The star Antares in Scorpio (the scorpion) means "anti-Aries, or, against Aries (the Ram/Lamb)." The name Antares corresponds, biblically, to the Greek word "antichrist" or the Hebrew word "anti-Messiah," which means "opposing the Messiah."

In Sagittarius, we can observe the prophetic implications of Messiah and his kingdom administration destroying the lying works of the antichrist spirit with his sharp arrows. These arrows represent descendants of the righteous who are taught the word of God by their parents, pastors, teachers, and mentors. With proper training, vision, and purpose, these "arrows" hit their target. When a child is reared in the ways of the Lord, he or she can be aimed, as an arrow, to do works of righteousness.

Parents must be students of God's word, in order to point their children in the right direction, as arrows in the hand of a warrior. Children of the Lord are created in the Messiah for good works (Eph 2:10). Through the truth of the word of God, followers of the Messiah can make disciples and destroy the lying works of the enemy in the lives of others.

As children and offspring of God (*bene ha Elohim*), we are given the same mandate as the captain of our salvation, Yeshua: "For this purpose the Son of God was manifested, that He might destroy the works of the devil" (1 John 3:8).

One pivotal purpose for godly children is to advance God's kingdom from generation to generation: "Behold, children are a heritage from the Lord, the fruit of the womb is a reward. Like arrows in the hand of a warrior, so are the children of one's youth. Happy is the man who has his quiver full of them; they shall not be ashamed, but shall speak with their enemies in the gate" (Ps 127:3–5).

Praise the Lord for his precious promise fulfilled in the homes of godly parents: "All your children shall be taught by the Lord, and great shall be the peace of your children" (Isa 54:13).

Conventional Star Name/Meaning	Name/Meaning Confirmed or Corrected
Al Nasl—(not included in E. W. Bullinger's book)	*Al Nasl*—included: offspring, arrowhead

Enclosed are derivatives and similar words in ancient languages which help clarify the core meaning of this constellation/star name.

Ancient Hebrew/ Aramaic	Phonetic Spelling	Summarized Meanings
Aramaic ܓܢܣܐ	G'eNSaA	offspring, kind

Ancient Gentile Language	Phonetic Spelling	Summarized Meanings
Arabic نَسْل	nasl	offspring, descendants
Arabic نَصْل	naṣl	arrowhead
Persian	nasl	offspring, progeny, lineage, pedigree
Persian	niṣāl	points of arrows; an arrow-head; spear-heads

Scripture References

Eph 2:10; 1 John 3:8; Ps 127:3–5; Isa 54:13

14

The Constellation: Lyra

THE NAME *LYRA* (PRONOUNCED ly-rae) is the Latin word for "lyre, harp." Rooted in the Greek word *lura*, lyre is related to the word *lyric*, derived from the Greek *lurikos*, or "singing with the lyre."

The constellation *Lyra* is a picture of the heavenly "music of the spheres" expressed through this instrument of praise. God fashioned Lyra as a symbolic reminder of joy and thanksgiving to Messiah in his victory over Satan—and paying the sin debt of mankind, on the cross, through the death, burial, and resurrection of his only begotten Son.

Lyra graces the northern celestial hemisphere next to Cygnus, the Swan, known as the Northern Cross. As a classical guitarist, Lyra is particularly significant for me. The chorus in my composition "The Cross and the Harp" relays the message of their union:

> The cross and the harp side by side
> In the sky, lift your eyes
> The cross and the harp night by night
> Oh, hear God's message of light

The Hebrew counterpart for the Latin word *lyre* is כִּנּוֹר/*kinnôwr* or *kinnor*. The *kinnor* was invented by Jubal, "Yubal," the great-great grandson of Adam's first son, Cain. "He was the father of all those who play the harp [כִּנּוֹר/*kinnôwr*] and flute" (Gen 4:21).

The *kinnor* was the first stringed instrument of its entire class. The *kinnor* has been called the national instrument of Israel. In modern Hebrew, the word *kinnor* refers to a violin.

The harp and lyre are mentioned numerous times in the Bible; the fundamental difference between the two has to do with the positioning of the strings. The strings of a harp enter directly into the hollow body of the instrument, whereas the strings of a lyre pass over the "bridge" of the instrument, transmitting vibrations from the strings to the body of the instrument—as does a modern-day guitar.

The *kinnor* is used specifically as an accompaniment to songs of cheerfulness, and for praise and worship to God.

As King David transported the ark of the covenant to Jerusalem, he danced before the Lord accompanied by musicians playing the *kinnor* and other instruments of praise: "Thus, all Israel brought up the Ark of the Covenant of the Lord with shouting and with the sound of the horn, with trumpets and with cymbals, making music with stringed instruments and harps [כִּנּוֹר/*kinnôwr*]" (1 Chr 15:28).

Psalmists were called and appointed to prophesy God's word on stringed instruments (כִּנּוֹר/*kinnôwr*) in the tabernacle of David.

> And the number of the skilled men performing their service was: Of the sons of Asaph: Zaccur, Joseph, Nethaniah, and Asharelah; the sons of Asaph were under the direction of Asaph, who prophesied according to the order of the king. Of Jeduthun, the sons of Jeduthun: Gedaliah, Zeri, Jeshaiah, Shimei, Hashabiah, and Mattithiah, six, under the direction of their father

Jeduthun, who prophesied with a harp [כִּנּוֹר/kinnôwr] to give thanks and to praise the Lord. (1 Chr 25:1–3)

According to Acts chapter 15, the apostle James (half-brother of Yeshua) explains that the tabernacle of David will be restored within the new covenant church—the spiritual Israel of God (Jew and Gentile, one in Messiah). This restoration will result in the salvation of the Jewish people and Gentiles turning to the God of Israel. This restoration, occurring over the last two thousand years, fulfills biblical prophecies relating to the unity of the body of Messiah—known as the "mystery" of the "One New Man," according to Ephesians 2:14; 4:13.

Indeed, Yeshua's high priestly prayer, as recorded in John chapter 17, is being answered through the unity of his spiritual body, the church. The apostle James's words are also being fulfilled in our generation: "And with this the words of the prophets agree, just as it is written: 'After this I will return and will rebuild the Tabernacle of David, which has fallen down; I will rebuild its ruins, and I will set it up; so that the rest of mankind may seek the Lord, even all the Gentiles who are called by My name, says the Lord who does all these things'" (Acts 15:16, 17).

The apostle James bases his message on the prophecy of Amos. In Amos chapter 9, God presents a two-fold prophecy regarding the restoration of the land of Israel and the restoration of the tabernacle of David (see Amos 9:11–15).

King David set the precedent for praise to the Lord by setting to music the psalms he wrote and played on the *kinnor*. When David played the *kinnor* (harp) before Saul, the evil spirit departed (1 Sam 16:14–23). David's musical connection with God expelled the forces of evil. Evil spirits do not have the power to remain in God's presence. Vibrations caused by hatred, confusion, fear, turmoil, envy, and strife cannot abide in the presence of the Lord.

One of the primary purposes for music is to prophesy the word of God. The Hebrew word for prophesy is נָבָא/nâbâ' (naw-baw'), meaning "a primitive root; to prophesy, i.e., speak (or sing) by inspiration (in prediction of, or simple discourse); to prophesy."

The Greek word for prophecy is *prophéteuó*, meaning "to foretell, tell forth; 'speak forth' in divinely empowered forthtelling or foretelling." To prophesy with the harp and other musical instruments is to declare God's word accompanied by a musical instrument. In short, praise and worship unto God is part of our spiritual DNA.

Praise can be employed, even by children in "*war*ship," as stated in the psalm of David: "Out of the mouth of babes and nursing infants You have ordained strength [עֹז/ *ʿôz*], because of Your enemies, that You may silence the enemy and the avenger" (Ps 8:2; Matt 21:16).

This Hebrew word for strength is עֹז/ *ʿôz*, meaning "boldness, loud, might, power, strength, strong." In fact, the Lord is our strength, song, and salvation: "The Lord is my strength [עֹז/ *ʿôz*] and song, and He has become my salvation" (Ps 118:14). Even our instruments can have God's anointed strength upon them: "and the Levites and the priests praised the Lord, day by day, singing to the Lord, accompanied by loud [עֹז/ *ʿôz*] instruments" (2 Chr 30:21).

We must be ever mindful as to whom and to what we listen. "Then Yeshua said to them, 'Take heed what you hear. With the same measure you use, it will be measured to you; and to you who hear, more will be given'" (Mark 4:24). We would do well to listen to godly music. Is the music to which we're listening "vibrating" on the same frequency as our Creator?

The apostle Paul describes the warfare in which we are engaged:

> Finally, my brethren, be strong in the Lord and in the power of His might. Put on the whole armor of God, that you may be able to stand against the wiles of the devil. For we do not wrestle against flesh and blood, but against principalities, against powers, against the rulers of the darkness of this age, against spiritual hosts of wickedness in the heavenly places. Therefore, take up the whole armor of God, that you may be able to withstand in the evil day, and having done all, to stand. (Eph 6:10–13)

"For we do not wrestle against flesh and blood" tells us our warfare is not with human beings—in and of themselves. Therefore, we must be aware of what we hear and to whom we listen lest we come under the influence of people who, knowingly or unknowingly, are tuned into the same "frequency" as Satan. The Greek word (in Eph 6:12) for "wrestle" is πάλη/*páleto* from πάλλω/*pállō*, meaning "to vibrate; wrestling—wrestle." Spiritual warfare has to do with to whom and to what we listen. The Hebrew word for listen, hear, and obey is שָׁמַע/*shâma* (Deut 6:4).

God desires that we "agree" with him, living our lives within the "frequency" of his holy, inspired word. Matthew's Gospel gives us insight into the word "agree," the root word for "symphony." The Greek word is συμφωνέω/*sumphóneó* and means "to agree together." Its root is from φωνή/*phōné*, meaning "a sound, a tone, of inanimate things, as musical instruments, to be harmonious, of the sound of uttered words."

"Again, I say to you that if two of you agree [*sumphóneó*] on earth concerning anything that they ask, it will be done for them by My Father in heaven. For where two or three are gathered together in My name, I am there in the midst of them" (Matt 18:19, 20).

There is power in the unity of the Holy Spirit. Psalm 133 says: "Behold, how good and how pleasant it is for brethren to dwell together in unity!" (Ps 133:1). On the day of Shavuot/Pentecost, "they were all with one accord in one place. And suddenly there came a sound from heaven, as of a rushing mighty wind, and it filled the whole house where they were sitting" (Acts 2:1, 2).

When the frequency of our heart is tuned to the master conductor of the universe our lives become beautiful, harmonious music to his name.

The apostle Paul exhorts us to know and experience the true purpose of music: "And do not be drunk with wine, in which is dissipation; but be filled with the Spirit, speaking to one another in psalms and hymns and spiritual songs, singing and making melody in your heart to the Lord" (Eph 5:18, 19).

The Greek word for psalm is *psalmos*, defined as "a striking (of musical strings), a psalm, scripture set to music." Originally, a psalm was sung and accompanied by a musician plucking a musical instrument (typically, the harp), particularly, the Hebrew Psalms.

The Hebrew word for psalm is זָמַר/*zâmar* (zaw-mar'), meaning "a primitive root (relating to the idea of striking with the fingers); properly, to touch the strings or parts of a musical instrument, play upon it; to make music, accompanied by the voice; hence to celebrate in song and music—give praise, sing forth praises, psalms."

The Creator is the inspiration of godly music. And music is one of the creative ways to praise and magnify his name and word.

The classic messianic psalm regarding the crucifixion of the Messiah (written in 1000 BC) says this about praise: "But You are holy, enthroned in the praises of Israel" (Ps 22:3). The Hebrew word "praises" is תְּהִלָּה/*tᵉhillâh*, meaning "laudation; specifically (concretely) a hymn: praise." The Hebrew word enthroned is יָשַׁב/*yâshab*, meaning "to dwell, remain, sit, abide, to marry." The kingship authority of God resides in his people who enthrone him in praise.

Through music we praise God for what he has done in our lives and we worship him for who he is. Worship to our Creator, Redeemer is the highest form of music.

The heavenly constellation Lyra reminds us that our focus should be on worshiping and praising the Lord.

There are seven primary Hebrew words for praise:

1. *Hâlal*/הָלַל—"to shine, to praise, boast, be boastful." "Praise [*hâlal*] the Lord! Praise [*hâlal*], O servants of the Lord, praise [*hâlal*] the name of the Lord!" (Ps 113:1).

2. *Yâdâh*/יָדָה—"to use (i.e., hold out) the hand; especially to revere or worship (with extended hands)." "Oh, that men would give thanks [*yâdâh*] to the Lord for His goodness, and for His wonderful works to the children of men!" (Ps 107:15). "I desire therefore that the men pray everywhere, lifting up holy hands, without wrath and doubting" (1 Tim 2:8).

3. *Tôwdâh*/תּוֹדָה—"an extension of the hand (by implication), avowal, or (usually) adoration; specifically, a choir of worshipers: confession, (sacrifice of) praise, thanks-giving." "Offer to God thanksgiving [*tôwdâh*], and pay your vows to the Most High" (Ps 50:14).

4. *Shâbach*/שָׁבַח—"to laud, praise, adulate, adore." "One generation shall praise [*shâbach*] Your works to another, and shall declare Your mighty acts" (Ps 145:4).

5. *Bârak*/בָּרַךְ—"to kneel; by implication to bless God (as an act of adoration)." "I will bless [*bârak*] the Lord at all times; His praise shall continually be in my mouth" (Ps 34:1).

6. *Zâmar*/זָמַר—"the idea of striking with the fingers; properly, to touch the strings or parts of a musical instrument." "Be exalted, O Lord, in Your own strength! We will sing and praise [*zâmar*] Your power" (Ps 21:13).

7. *T^ehillâh*/תְּהִלָּה—"laudation; specifically (concretely) a hymn: praise." "But You are holy, enthroned in the praises [*t^ehillâh*] of Israel" (Ps 22:3).

Today, we are to let the word of Messiah dwell in us richly, so his body will grow: "Let the word of Christ dwell in you richly in all wisdom, teaching and admonishing one another in psalms and hymns and spiritual songs, singing with grace in your hearts to the Lord" (Col 3:16).

There are three major star names represented in Lyra, contributing to its message.

1. The star name Vega: to "extol, praise."

2. The star name Sheliak or Shelyak: "sprout/branch, sent, sword."

3. The star name Sulaphat: "tortoise"; according to ancient Greek tra-
    dition, the lyre was made from a turtle (tortoise) shell.

These star names are defined in this chapter under the title of their
names.

Lyra's brightest star, Vega, is one of the most recognizable stars in
the night sky. My triumphant song of the harp truly proclaims "Yeshua is
highly exalted."

He Is Highly Exalted

(Phil 2:5–11)
By Bruce J. Patterson

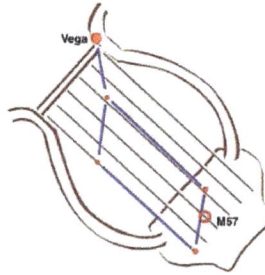

Image by the University of Texas McDonald Observatory, https://stardate.org.

Verse 1
The song of songs in preeminent gladness
Played on the harp of heaven in brightness
Placed among the stars in celestial sphere
His triumphant song we can hear

Chorus
Yeshua, he is highly exalted
Yeshua, he is highly exalted

Verse 2
The highest name in the universe displayed
In pictorial prophecy we may read

And join the joyful song the holy angels sing
Halleluyah to our King

Chorus

Verse 3
Praise to the conqueror
Praise him with the harp
Praise to Yeshua
He is highly exalted

Constellation Name	Translation
Lyra (Latin)	lyre, harp
Lyrê (Greek)	the lyre
Sheliak (Arabic)	the lyre
Al-Qitharah (Arabic)	the lyre
tiān qín zuò (Chinese)	the celestial zither constellation (zither: a stringed instrument)

Enclosed are derivatives and similar words in ancient languages which help clarify the core meaning of this constellation/star name.

Ancient Hebrew/ Aramaic	Phonetic Spelling	Summarized Meaning
Hebrew כִּנּוֹר	kinnôwr	lyre, harp
Aramaic ܩܺܝܬ݂ܳܪܳܐ	QiYT,aRaA	harp

Ancient Gentile Language	Phonetic Spelling	Summarized Meaning
Latin	lyrae	lyre, harp
Greek	lýra	lyre, harp

Scripture References

Gen 4:21; 1 Chr 15:28, 25:1–3; Acts 15:16–17; Amos 9:11–15; 1 Sam 16:14–23; Eph 6:10–13, 5:18–19; Psa 49:4, 150:3; 1 Kgs 10:12; 1 Chr 15:16; Rev 14:2

The Stars

Vega

THE STAR NAME VEGA is a possible derivative of the Hebrew word שָׂגָא/ sâgâ, meaning "to grow, i.e., (causatively) to enlarge, (figuratively) laud— increase, magnify, extol with praise, laud." E. W. Bullinger in *The Witness of the Stars* defines Vega as "he shall be exalted."[1] A similar pronunciation and meaning can be found in the selective Gentile languages (see chart).

An example of the Hebrew שָׂגָא/sâgâ is found in the book of Job, where Elihu proclaims God's majesty: "Remember to magnify [שָׂגָא/sâgâ] His work, of which men have sung" (Job 36:24).

Sâgâ is related to the Hebrew שָׂגַב/sâgab, meaning "defend, exalt, be excellent, (be, set on) high, lofty, be safe, be too strong, exaltation." *Sâgab* occurs twenty times in the Hebrew Bible.

Isaiah's ancient messianic prophecy uses this word in reference to the reign of Messiah: "And the loftiness of man shall be bowed down, and the haughtiness of men shall be made low: and the LORD alone shall be exalted [שָׂגַב/sâgab] in that day" (Isa 2:11).

A word similar to *sâgab* (גָּאָה/gâ'âh) is found in the song of Moses: "Then Moses and the children of Israel sang this song to the Lord, and spoke, saying: 'I will sing to the Lord, for He has triumphed [גָּאָה/gâ'âh] gloriously! The horse and its rider He has thrown into the sea!'" (Exod 15:1). The Hebrew גָּאָה/gâ'âh means "to rise up, be exalted in triumph, be majestic."

Vega could be affiliated with the Persian and Arabic word *wāqi*, meaning "falling; a bird descending from the air." Vega could also be affiliated with the Persian *waqah*, meaning "listening to, obeying." This Persian word may be related to the ancient Hebrew יָקֶה/yâqeh, which means

---

1. Bullinger, *Witness of the Stars*, 56.

"to obey; obedient; blameless—gathering" and is used in reference to a person's name in Prov 30:1. It is a derivative of יִקָּהָה/*yiqqâhâh*, meaning "obedience." This word is used in reference to the coming Lion of the tribe of Judah, Yeshua, the Messiah: "The scepter shall not depart from Judah, nor a lawgiver from between his feet, until Shiloh comes; and to Him shall be the obedience [יִקָּהָה/*yiqqâhâh*] of the people" (Gen 49:10).

While we do not know the origin of the name "Vega," I include possible options in my Vega definition chart. Occasionally, it's challenging to decipher one- or two-syllable words which have similar phonetic pronunciations but different meanings in ancient languages.

Vega is the fifth brightest star in the night sky, and the second brightest star in the northern celestial hemisphere subsequent to Arcturus. According to astronomical math, Vega is projected to become the northern polestar—the star which determines the center of the celestial sphere in the year 13727—as gloriously expected!

Due to the precession of the equinoxes, the star Errai, which means "the shepherd," in Cepheus (symbolizing Messiah the King), will become the northern polestar in about one thousand years (around AD 3000), only to become more centralized in approximately AD 4000.

Yeshua has been reigning from his throne on high for nearly two thousand years. The Bible says, "He must reign until He has put all enemies under His feet" (1 Cor 15:25).

Epic Discovery

On resurrection morning (April 29, AD 31), Vega crosses the meridian with Lyra in zenith position at 3:12 a.m.

Vega could be seen seven weeks after the resurrection in zenith position on the Feast of Shavuot/Pentecost, Sunday, June 17, AD 31, precisely at 12:00 midnight. The heavens truly declare Yeshua is highly exalted! Vega, in Lyra (the Harp), amplifies the victory song of Messiah who has accomplished for us so great a salvation.

Yeshua is highly exalted and deserves our highest praise and worship.

Conventional Star Name/Meaning	Name/Meaning Confirmed or Corrected
Vega—he shall be exalted	confirmed

Enclosed are derivatives and similar words in ancient languages which help clarify the core meaning of this constellation/star name.

Ancient Hebrew/ Aramaic	Phonetic Spelling	Summarized Meanings
Hebrew שָׂגַב	*sâgab*	defend, exalt, be excellent, (be, set on) high, lofty, be safe, be too strong) exaltation
Aramaic ܙܐ	*ZaH*	exalted
Hebrew שָׂגָא	*sâgâ*	a primitive root; to grow, i.e., (causatively) to enlarge, (figuratively) laud—increase, magnify. Extol with praise, laud (Job 36:24)
Aramaic ܐܣܓܐ	*sᵉgâ*	corresponding to Hebrew; שָׂגָא/*sâgâ*
Hebrew גָּאָה	*gâ'âh*	to rise up, be exalted in triumph, be majestic
Hebrew וְיִגְאֶה	*wə·yig̱·'eh*	should [my head] be lifted
Hebrew יִגְאֶה	*yig̱-'eh*	grow up (tall), flourish (Job 8:11)
Hebrew יְקֶה	*yâqeh*	to obey; obedient; blameless
Hebrew יִיקָּהָה	*yiqqâhâh*	obedience—gathering, to obey

Ancient Gentile Language	Phonetic Spelling	Summarized Meanings
Sanskrit	*jaya*	triumph
Assyrian	*za: ka*	victory
Arabic	*wāqi*	falling
Arabic	*waqa*	to come down, fall
Arabic	*waq'a*	battle
Arabic	*wajīh*	sound, good, valid

Ancient Gentile Language	Phonetic Spelling	Summarized Meanings
Persian	*wāqi*	falling, a bird descending from the air
Persian	*waqā*	that by which anything is kept or preserved; preserving, guarding; a shield
Persian	*waqah*	listening to, obeying
Persian	*fa-egh*	excellent, superior, superfine
Persian	*faeghah*	high(est) or great(est), to excel; to attain eminence or superiority
Sumerian	*a-zi-ga*	rising waters ("water" + "to rise up")
Assyrian	*sia: qa*	going up, ascending
Assyrian	*e ' sa: qa*	to ascend, to rise
Akkadian	*Šaqû*	(astronomy) to raise up; to exalt; to bring, travel upstream, to exalt, elevate; to praise, to rise
Sumerian	*še-ga/ge*	favorite; to be obedient; to obey; to agree

Scripture References

Job 36:24; Exod 15:1; Isa 2:11; 1 Cor 15:25

Sheliak or Shelyak

The star name Sheliak (pronounced SHEL-ee-yak) may be associated with the Greek word *chelóna*, which means "tortoise, turtle." A tortoise shell was often used as the "body" of the lyre, an ancient type of harp.

The *chelys* (Greek: χέλυς, Latin: *testudo*), rooted in the ancient Greek word χέλῡς (*khélūs*, "tortoise"), is a string musical instrument, the common lyre of the ancient Greeks having a convex back of tortoise shell, or wood, shaped like a shell.

Lyra's traditional name is Sheliak, from الشلیاق/*šiliyāq*, the Arabic name of the constellation. The Bayer designation for Sheliak was given by the German astronomer Johann Bayer in the 1603 publication of his star atlas, called *Uranometria*.[2]

Shalyāk, the Persian name for the constellation Lyra, is a derivative of *shāshak*, which means "instrument of music." Another Persian derivative of Shalyāk is the word *shalīḵẖ*, or "sound, noise." Yet another possible derivative is *shalīḵẖā*, meaning "the companions of Jesus; a Christian."

The Hebrew language is older than Arabic and Persian. Therefore, it's possible that the Arabic and Persian derivatives are an offshoot of the ancient Hebrew שֶׁלַח/*Shelach*, meaning "to send, sent, a shoot of growth; sprout, branch, a sword." Furthermore, the star Sheliak may be related to the Hebrew word שָׁלִישׁ/*shâlîysh*, meaning a "triangle" (the musical instrument), or the three-stringed lute.

An example of שָׁלִישׁ/*shâlîysh* can be found in the book of 1 Samuel: "And it came to pass as they came, when David was returned from the slaughter of the Philistine, that the women came out of all cities of Israel, singing and dancing, to meet king Saul, with tabrets, with joy, and with instruments [שָׁלִישׁ/*shâlîysh*] of music" (1 Sam 18:6). Another example is found in reference to captains: "But of the children of Israel did Solomon make no servants for his work; but they were men of war, and chief of his captains [שָׁלִישׁ/*shâlîysh*], and captains of his chariots and horsemen" (1 Kgs 9:22).

The star Sheliak, in the constellation Lyra, depicts the victory song of Yeshua, the captain of our salvation. All praise belongs to him, the captain of the Lord of Hosts!

Conventional Star Name/Meaning	Name/Meaning Confirmed or Corrected
Sheliak or *Shelyak*—eagle	corrected: tortoise, instrument of music

Enclosed are derivatives and similar words in ancient languages which help clarify the core meaning of this constellation/star name.

2. Definitions.net, "Beta Lyrae."

Ancient Hebrew/ Aramaic	Phonetic Spelling	Summarized Meaning
Hebrew שָׁלִישׁ	shâlîysh	a triangle (or perhaps rather three-stringed lute); also (as an indefinite, great quantity), a three-fold measure (perhaps a treble ephah); also (as an officer) a general of the third rank (upward, i.e., the highest), captain, instrument of music, (great) lord, (great) measure, prince, three
Hebrew שְׁלִישִׁי	sheliyshiy	(third) from: שָׁלוֹשׁ/ shâlôwsh (three, triad)
Hebrew שֶׁלַח	shelach	a sprout, a weapon-sword Salah or Shelah = "sprout" Son of Arphaxad and father of Eber Salah, son of Arphaxad, and grandfather of Abraham
Hebrew שָׁלַח	shâlach	to send away, for, or out (in a great variety of applications)—× any wise, appoint
Aramaic	shelach	to send (Dan 5:24, 6:22)

Ancient Gentile Language	Phonetic Spelling	Summarized Meaning
Greek	chelőna	tortoise, turtle
Greek	chelys	tortoise, tortoise-like
Persian	shāshak	instrument of music
Persian	Shalyāk	the constellation of the Lyre
Persian	shash-tā	a kind of lute with six strings
Persian	shalīkhā	the companions of Jesus; a Christian
Persian	shalīkh	sound, noise
Akkadian	shalash	three

Ancient Gentile Language	Phonetic Spelling	Summarized Meaning
Phoenician	shlsh	three
Assyrian	shlakh	to send, to dispatch
Amharic/Ethiopian	lake	send

Scripture References

1 Sam 18:6; 1 Kgs 9:22

Sulaphat

The star name Sulaphat (pronounced SOOL-a-faht) originates in the Arabic سُلَحْفاة/sulaḥfat, meaning "tortoise." Sulaphat relates to the type of lyre made from a tortoise shell.

A classical lyre has a hollow body or "sound-chest" (also known as soundbox or resonator), which, in ancient Greek tradition, was made from tortoise shell. The lyre was the most important and well-known instrument in the Greek world.

The Hebrew kinnor was originally a box lyre, which is mentioned in the Psalms. The star name "Sulaphat" confirms that the constellation Lyra represents the musical instrument lyre; however, there may be an additional meaning of the word sulaphat.

Based on the first two syllables of the word sulaphat, sula may be a derivative of the Arabic word salām, meaning "peace" from the ancient Hebrew שׁוּלה/shula, meaning "peaceful," from שָׁלוֹם/shalom (peace).

Those who know God through their relationship with his son, Yeshua Hamashiach (Jesus, the Messiah), can rest assured that God is the God of our peace, shalom, which in the original Hebrew means "wholeness, nothing missing or lacking, complete well-being."

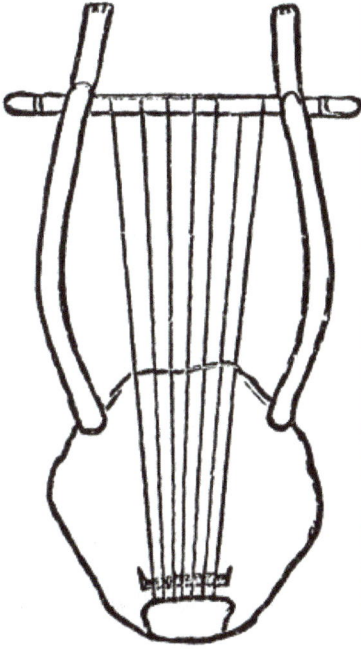

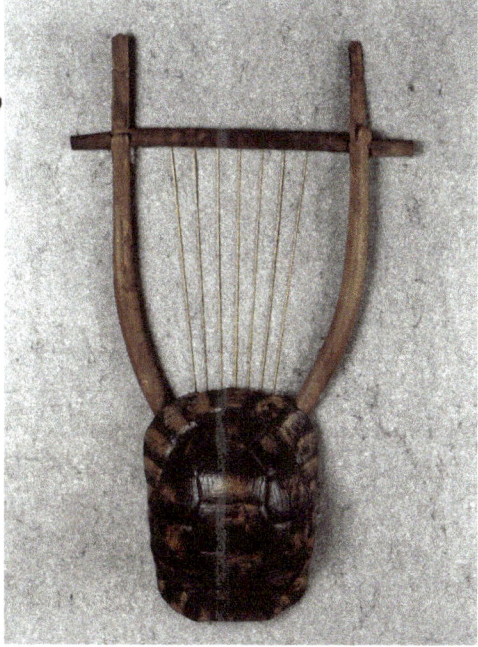

LEFT: Ancient Greek *chelys* or lyre sketch, Florida Center for Instructional Technology Innovative Education, University of South Florida. RIGHT: Lyre, Classical Greek, Athens, the Trustees of the British Museum.

The fundamental difference between a harp and a lyre has to do with the positioning of the strings. The strings of the harp enter directly into the hollow body of the instrument, whereas the strings of a lyre pass over a bridge, which transmits the vibrations of the strings to the body of the instrument—just as a modern guitar.

The harp and lyre are plucked with the fingers. The word *psalm* means to "pluck with the fingers on a string instrument." The Hebrew word for psalm, זָמַר/*zâmar*, and the Greek word for psalm, ψαλμός/*psalmos*, both mean "striking a musical instrument with the fingers."

The Spirit of Yᵉhôvâh, who inspired Moses, the Prophets, and the Apostles to write the holy Scriptures, is still at work today. The Lord simply needs yielded vessels through whom he can communicate his thoughts of wisdom and counsel based on the canon of Scripture he has provided. The Spirit of Yᵉhôvâh, who inspired the true meaning of the constellations, is at work today. Both the Bible and the Mazzaroth are of the same author, Yᵉhôvâh Almighty.

As followers of Yeshua, let us heed the advice of the angel who spoke to the apostle John in the book of Revelation: "Worship God! For the testimony of Jesus is the spirit of prophecy" (Rev 19:10).

Conventional Star Name/Meaning	Name/Meaning Confirmed or Corrected
Sulaphat—springing up, ascending	corrected: tortoise (Arabic)

Enclosed are derivatives and similar words in ancient languages which help clarify the core meaning of this constellation/star name.

Ancient Hebrew/ Aramaic	Phonetic Spelling	Summarized Meaning
Hebrew שׁוּלה	*shula*	(peaceful) from: שָׁלוֹם/ *shalom* (peace)
Hebrew שָׁפַט	*shaphat*	judge
Hebrew שׁוּלַמִּית	*shulammith*	(from *shalam*) meaning peaceful
Aramaic ܫܠܡܐ	*SHaLMaA*	whole, entire, perfect, peace, salute, salutation
Hebrew שָׁלַף	*shalaph*	to draw out or off
Aramaic ܫܡܛ	*SHMaT*	draw

Ancient Gentile Language	Phonetic Spelling	Summarized Meaning
Arabic سُلَحْفَاة	*sulaḥfāï*	tortoise, turtle
Egyptian Arabic	*sulhifaea*	tortoise
Assyrian	*sa lu: ʿpi: ta*	a tortoise
Akkadian	*Šeleppūtu*	a tortoise
Persian	*shailūna*	a tortoise
Akkadian	*šeleppūtu*	turtle, tortoise
Arabic	*salām*	peace
Persian	*salām*	peace

Ancient Gentile Language	Phonetic Spelling	Summarized Meaning
Turkish	*selam*	peace
Assyrian	*shla: ma*	peace

Scripture References

Song 6:13; Isa 53:5, 54:10; Pss 9:8, 50:6; Rev 19:10

15

The Constellation: Ara

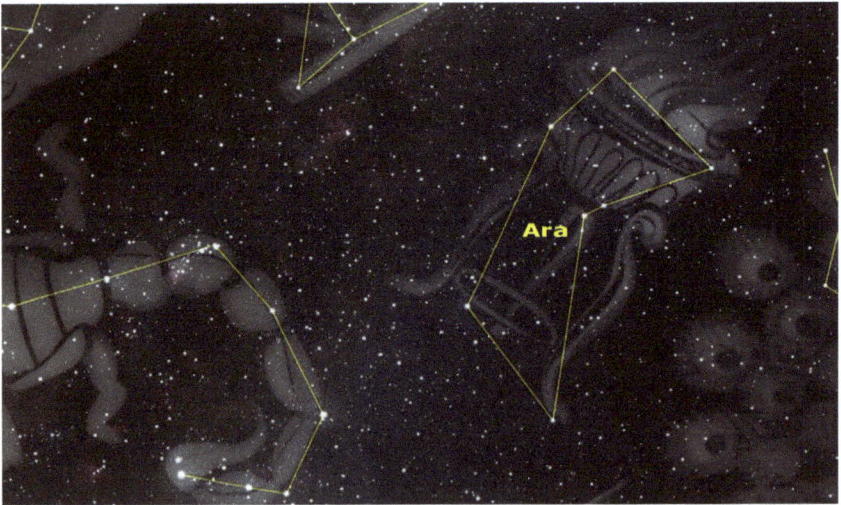

THE WORD *ĀRA* (PRONOUNCED AY-rah) is Latin for "a structure for sacrifice; altar." The constellation Ara portrays the old covenant sacrificial
atonement system which foreshadows the new covenant sacrifice and
atonement of Jesus, Yeshua, the Lamb of God.

Although not stated by name, the constellation Ara, the Altar, may
be one of the constellations in the "chambers of the south," as cited in
the book of Job: "He commands the sun, and it does not rise; He seals
off the stars; He alone spreads out the heavens, and treads on the waves
of the sea; He made the Bear, Orion, and the Pleiades, and *the chambers*

of the south; He does great things past finding out, yes, wonders without number" (Job 9:7–10).

Positioned behind Scorpio, the Scorpion, Ara is depicted as having been overturned by the scorpion. In essence, the "overturned" constellation, Ara, is symbolic of the enemy perverting justice and attempting to overthrow Christianity.

Antiochus IV Epiphanes, King of Syria, captured Jerusalem in 167 BC, desecrating the Jewish temple by sacrificing a pig on an altar dedicated to a Syrian god. Desecrating the Jewish temple is a picture of the enemy perverting and overthrowing aspects of Christianity.

The book of Daniel refers to the desecration of the temple as "the Abomination of Desolation" (Dan 12:11). Seeking to prohibit Judaism, in an attempt to "Hellenize" the Jews (to be made Greek, in form or character), Antiochus banned Jewish religious policies and practices, ordering copies of the Torah be burned.[1]

The attack against Judeo-Christianity continues to propagate a world view of man-made religion, legalism, socialism, communism and marxism throughout the centuries.

God yearns for his saints to possess his kingdom and fill the earth with the knowledge of the Lord as the waters cover the sea (Hab 2:14; Dan 7:22). The seed of the serpent, however, uses religion, power, and money to oppose and stifle the work of God in an effort to prevent his saints from possessing the kingdom of God.

The ongoing war between the seed of the serpent (Satan) and the Seed of the Woman (the Redeemed) continues to this day. The master of deception attempts to tilt the scales of justice (as displayed by Scorpio, the Scorpion) through religion and the traditions of man, thus making void the commandments of God. The enemy uses agents of Adam to tilt the scales in fraudulent governmental elections. The enemy interferes with the saint's sacrificial service to the Lord. The enemy attempts to "alter the altar" of the saint's service to the Lord.

The enemy wages war against the saints based on lies. The saints wage war against spiritual darkness based on truth. The saints' warfare is not against flesh and blood, but against serpentine powers of darkness (see Eph 6:12).

---

1. Josephus, *Works of Josephus,* "Antiquities of the Jews," 12.5.4.

Altars are a place of sacrifice. The old covenant brazen altar typifies the cross of Calvary. Yeshua fulfilled the ultimate purpose of the altar. Today, we have an option. Will we sacrifice our lives on the altar of the Lord for godly things, or will we sacrifice our lives on a worldly altar for worldly things? We can choose to give the Lord a "sacrifice of praise, to do good and give," each day (Hebrew 13:15, 16).

The Hebrew word "altar" in the Bible, מִזְבֵּחַ/*mizbêach*, pronounced miz-bay'-akh, has its origin in the word זָבַח/*zâbach*, zaw-bakh', meaning "to slaughter an animal (in sacrifice): to kill, offer, (do) sacrifice, slay."

An altar was constructed by the patriarch Noah after the flood. Abraham built an altar upon which to offer Isaac, but the altar was used to offer a sacrificial ram (a picture God's Son) in his place. Jacob built an altar at Bethel. Moses built an altar at the foot of Mount Sinai. The nation of Israel built the brazen altar inside the courtyard of the tabernacle, and later the brazen altar of Solomon's Temple. And Elijah built an altar in the name of the Lord.

Only "pure" sacrifices (without spot or blemish) were to be offered on the brazen altar; the brazen, or "brass," altar, which typifies judgment, served as a shadow of the cross upon which the pure, sinless Lamb of God was judged for our sins.

Our redemption is not based on religion but reality: "Knowing you were not redeemed with corruptible things, like silver or gold, from your aimless conduct received by tradition from your fathers, with the precious blood of Christ, as of a lamb without blemish and without spot. He indeed was foreordained before the foundation of the world, but was manifest in these last times for you" (1 Pet 1:18–20).

All who receive the Lord Jesus and his sacrificial atonement on their behalf are pardoned for their sins. God's people are also identified with the crucifixion of Yeshua on the (altar) cross. As the apostle Paul said: "I have been crucified with Christ; it is no longer I who live, but Christ lives in me; and the life which I now live in the flesh I live by faith in the Son of God, who loved me and gave Himself for me" (Gal 2:20). Therefore, we are able to present our bodies as a living sacrifice to God (Rom 12:1, 2).

The book of Hebrews states: "We have an altar from which those who serve the tabernacle have no right to eat. For the bodies of those animals, whose blood is brought into the sanctuary by the high priest for sin, are burned outside the camp. Therefore Jesus also, that He might sanctify the people with His own blood, suffered outside the gate" (Heb 13:10, 12).

Yeshua's death, burial and resurrection fulfilled and finalized the old covenant Levitical priesthood. The Lord instituted the old covenant priesthood to provide *temporary* atonement for the sins of the nation of Israel which was *permanently* fulfilled by the Messiah.

God provides eternal atonement and remission of sins for his redeemed through the blood of the everlasting covenant, cut in the body of the Lamb of God. This redemptive act includes all who were atoned for during the old covenant dispensation.

The thirteenth chapter of the book of Hebrews describes priests eating the old covenant animal sacrifices. Today, Jews and Gentiles are invited to "spiritually feed" on Yeshua, whose flesh gives life to the world (John 6:53–58). This spiritual reality is communicated through the Lord's Passover, also called Communion and the Lord's Supper.

Yeshua, Jesus, *is* the Passover Lamb—the Lamb of God who takes away the sin of the world (John 1:29). His fleshly sacrifice enables believers to offer spiritual sacrifices (Heb 10:19, 20).

As the holy priesthood of the Lord God of Israel, we have an altar upon which we may offer "sacrifices," sacrifices of praise, doing good, giving, labors of love, ministering to the saints (Heb 13).

We have an altar within the context of the new covenant temple: "Coming to Him as to a living stone, rejected by men, but chosen by God and precious, you also, as living stones, are being built up a spiritual house, a holy priesthood, to offer up spiritual sacrifices acceptable to God through Jesus Christ" (1 Pet 2:4, 5).

"But you are a chosen generation, a royal priesthood, a holy nation, His own special people, that you may proclaim the praises of Him who called you out of darkness into His marvelous light" (1 Pet 2:9).

As a new covenant priesthood, we can understand and utilize the purpose of God's altar: "Oh, send out Your light and Your truth! Let them lead me; let them bring me to Your holy hill and to Your tabernacle. Then I will go to the altar of God, to God my exceeding joy; and on the harp I will praise You, O God, my God" (Ps 43:3, 4).

The heavens are "declaring" through Ara, the Altar, God's ultimate expression of love. Though the enemy attempts to distort the gospel in the stars and the pages of Scripture, the enemy cannot alter the word that has gone out of the Lord's lips (Ps 89:34).

For God so loved the world that He gave His only begotten Son,
that whoever believes in Him should not perish but have everlasting life. (John 3:16)

Constellation Name	Translation
Ara (Latin)	the altar
Thytêrios (Greek)	the altar
Al-Mijmarah (Arabic)	the censer
Al Mugamra (Arabic)	the completing
Tiān tán zuò (Chinese)	the heaven altar constellation

Enclosed are derivatives and similar words in ancient languages which help clarify the core meaning of this constellation/star name.

Ancient Hebrew/ Aramaic	Phonetic Spelling	Summarized Meaning
Hebrew מִזְבֵּחַ	mizbeach	an altar, from זָבַח/zâbach, meaning "to slaughter an animal (usually in sacrifice):—kill, offer, (do) sacrifice, slay

Ancient Gentile Language	Phonetic Spelling	Summarized Meaning
Latin	Āra	a structure for sacrifice, altar
Sanskrit	ara	altar
Phoenician	mzbh	altar
Arabic	taghayyara	altar
Akkadian	gamāru	end, completion

Scripture References

Job 9:7–10; Dan 12:11; Hab 2:14; 1 Pet 1:18–20; Gal 2:20; Heb 13:10–12; John 6:53–58; Rom 12:1; 1 Pet 2:4–5, 2:9; Ps 89:34; John 3:16; Exod 27:1; 2 Chr 1:6; Ps 118:27; Ezek 43:15

16

The Constellation: Draco

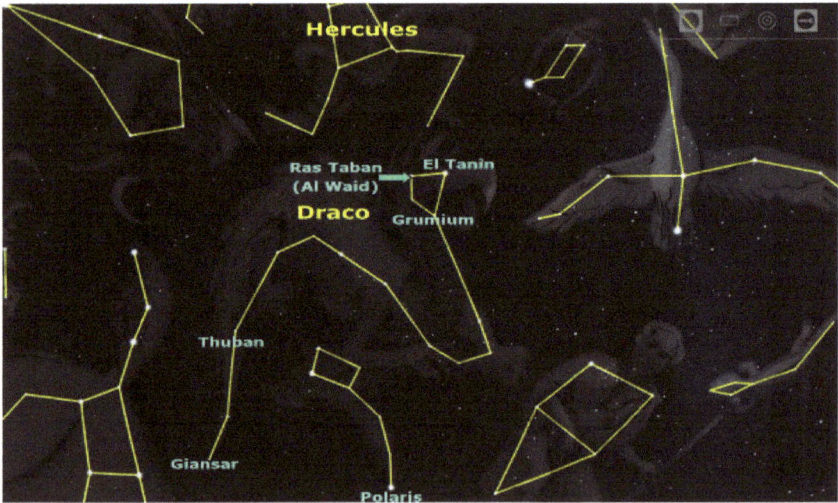

THE NAME *DRACO* IS the Latin word for "dragon." Draco is phonetically similar to the Hebrew word *Darak*, which means "to tread upon."

The corresponding biblical Hebrew name for this constellation, עֲקַלָּתוֹן נָחָשׁ לִוְיָתָן/*Livyathan Aqallathon Nachash*, means "Leviathan, the twisted serpent" (Isa 27:1). The Hebrew word לִוְיָתָן/*Livyathan* means "serpent, sea monster, or dragon," from the root word לָוָה/*lâvâh*, which means "to entwine by implication; to unite, to join." The Hebrew word עֲקַלָּתוֹן/*Aqallathon* means "crooked, twisted," from the root word עָקַל/

aqal, which means "to bend, twist, pervert, crooked justice." The word *Nachash* means "serpent" in Hebrew.

Draco, the Dragon (symbolizing Satan) is alluded to in the twelfth chapter of the book of Revelation. Draco is positioned in the northern region of the sky. This exalted position was claimed, symbolically, by Satan, after Adam and Eve sinned against God and lost their position of authority.

The constellation Draco, the Dragon, faces Virgo, the Virgin, anticipating, as it were, the birth of the Son of God. The earthly manifestation of Draco occurred when King Herod—in an attempt to "devour" the Virgin's child, "Yeshua, son of Miriam"—issued a decree calling for male babies two years of age and younger to be killed.

Draco, the Dragon (the twisted serpent), is winding around Ursa Minor (Dover, the Lesser Sheepfold). Dover represents the saint's inheritance in Messiah. The dragon's posture indicates the enemy's scheme: to steal, kill, and destroy God's people. Yet, he cannot reach God's elect.

Draco's brightest star, Thuban, or "snake" in Arabic, was the "polestar" (central pole position of the celestial sphere) four to five thousand years ago. Scripturally, Thuban represents Satan, the god and prince of this world. Satan became the god of this world by means of transferred authority. But Satan is no longer the god of God's redeemed (2 Cor 4:3, 4; John 14:30).

Thuban has long since been displaced by Polaris in Ursa Minor (see my explanation of the star Thuban in this chapter). Polaris represents restored authority.

Today, Polaris constitutes the center of the celestial sky. Hence, from Earth's perspective, everything revolves around Polaris, the polestar. The dominion Adam and Eve lost to the devil is restored by Yeshua, the last Adam. Ursa Minor (Dover, the Lesser Sheepfold) is a picture of the redeemed who reign with the Messiah in his heavenly kingdom—far above the enemies of God.

The head of Draco is depicted as being crushed by Hercules. The corresponding Hebrew word for Hercules, *Gibbôr*, means "strong, or mighty."

The true *Gibbôr* represents Yeshua, the Messiah of Israel. When the Messiah died on the cross, he thrust upward to crush the head of the serpent/dragon (authority) as stated in the ancient prophecy: "And I will put enmity between you and the woman, and between your seed and her Seed; He shall bruise your head, and you shall bruise His heel" (Gen 3:15).

God showcases the constellation Hercules crushing the head of Draco during the annual spring feasts of the Lord. Could this celestial scene tell the story of the first messianic prophecy (Gen 3:15) given by the Lord in the Bible? The answer is yes! The Messiah delivered us from the power of darkness and trampled Satan under his feet.

As previously stated, the Latin word Draco is phonetically similar to the Hebrew *Darak*, which means "to tread upon." *Darak* is fulfilled by the saints who exercise their God given authority according to the classic biblical promise: "You shall tread upon the lion and adder: the young lion and the dragon shall you trample [*darak*] under feet" (Ps 91:13).

Do you wonder why there is evil violence, immorality, calamity, tragedy, misery, pain, distress, suffering, sickness, disease, entropy, and death in this world—the origin of which continues taking its toll on humanity? Is there someone responsible for initiating evil on planet Earth? Would such a one be considered an enemy to humanity? These questions are answered in one word: Lūcifer.

The word Lūcifer is from the Latin *lux* and *ferre* meaning "light bringer." It is a transliteration of the Hebrew word, /הֵילֵל *Helel*, meaning "shining one." Helel is mentioned by name in the book of Isaiah:

> How you are fallen from heaven,
> O Lucifer, son of the morning!
> How you are cut down to the ground,
> You who weakened the nations!
> For you have said in your heart:
> "I will ascend into heaven,
> I will exalt my throne above the stars of God;
> I will also sit on the mount of the congregation
> On the farthest sides of the north;
> I will ascend above the heights of the clouds,
> I will be like the Most High God."
> Yet you shall be brought down to Sheol,
> To the lowest depths of the Pit.
> Those who see you will gaze at you,
> And consider you, saying:
> "Is this the man who made the earth tremble,
> Who shook kingdoms,
> Who made the world as a wilderness
> And destroyed its cities,
> Who did not open the house of his prisoners?" (Isa 14:12–17)

Isaiah gives us a glimpse into what motivated Helel to incite an attempted overthrow of the government of Yᵉhôvâh, the Creator. The root of rebellion is revealed in the heart of Helel through pride and self-will driven by envy and jealousy. Helel desired to be like God in position and power. This is why the constellation Draco is positioned in the *farthest sides of the north*. It exemplifies the exalted desire of Helel.

The prophet Ezekiel describes the origin of Helel:

> You were the seal of perfection,
> Full of wisdom and perfect in beauty.
> You were in Eden, the garden of God;
> Every precious stone was your covering:
> The sardius, topaz, and diamond,
> Beryl, onyx, and jasper,
> Sapphire, turquoise, and emerald with gold.
> The workmanship of your timbrels and pipes
> Was prepared for you on the day you were created.
>
> You were the anointed cherub who covers;
> I established you;
> You were on the holy mountain of God;
> You walked back and forth in the midst of fiery stones.
> You were perfect in your ways from the day you were created,
> Till iniquity was found in you.
>
> By the abundance of your trading
> You became filled with violence within,
> And you sinned;
> Therefore I cast you as a profane thing
> Out of the mountain of God;
> And I destroyed you, O covering cherub,
> From the midst of the fiery stones.
>
> Your heart was lifted up because of your beauty;
> You corrupted your wisdom for the sake of your splendor;
> I cast you to the ground,
> I laid you before kings,
> That they might gaze at you.
>
> You defiled your sanctuaries
> By the multitude of your iniquities,
> By the iniquity of your trading;
> Therefore I brought fire from your midst;

It devoured you,
And I turned you to ashes upon the earth
In the sight of all who saw you.
All who knew you among the peoples are astonished at you;
You have become a horror,
And shall be no more forever. (Ezek 28:12–19)

From the book of Ezekiel, we learn that Helel was once a beautifully created covering cherub, an order of angelic beings. He was created by the Almighty, the seal of perfection. Helel was a musician of the highest order. God built musical instruments within Helel. He occupied a high position in the presence of the Almighty and yet . . . he wanted to be higher. Following Helel's fall he became the father of lies (see John 8:34).

Ezekiel says: *You were in Eden, the garden of God.* We do not know how long Helel was in Eden before Adam and Eve were placed in that garden by God. Regardless of the timeline, Helel was lurking when Adam and Eve occupied the garden of Eden.

The Bible is clear as to the origin of sin: "He who sins is of the devil, for the devil has sinned from the beginning. For this purpose the Son of God was manifested, that He might destroy the works of the devil" (1 John 3:8).

Satan introduced deception to our first parents, Adam and Eve (the head of the human race). The devil tempted them to sin against God, so they would lose their headship dominion. Eve was deceived by the devil. If Adam had not succumbed to Eve's suggestion to eat of the forbidden fruit, the poison of sin would not have entered him. Adam sinned with his eyes wide open and broke the divine order. Only through the Messiah (the last Adam) can this divine order be restored.

"For Adam was formed first, then Eve. And Adam was not deceived, but the woman being deceived, fell into transgression" (1 Tim 2:13, 14).

It is important that we hear from the Almighty himself as to the origin of evil: "That they may know from the rising of the sun to its setting that there is none besides Me. I am the Lord, and there is no other; I form the light and create darkness, I make peace and create calamity [evil]; I, the Lord, do all these things" (Isa 45:6–7).

The Hebrew word for *evil* is רַע/*ra'*, meaning "bad, evil, adversity." Yeshua says in the sermon on the mount, *"deliver us from evil."* The Greek word for evil is *ponéros*, meaning "toilsome anguish, calamitous, diseased, bad, full of labors, annoyances, hardships."

How did God create evil? Considering that darkness exists where there is no light, and evil exists in the absence of good, we can surmise God created light and darkness, good and evil.

Before God created man, darkness and chaos covered planet Earth presumably due to the fall of Helel (Gen 1:1, 2; Isa 45:18). Genesis 1:1, 2 says: "In the beginning God created the heavens and the earth. The earth was without form, and void; and darkness was on the face of the deep."

The Hebrew word for formless is תֹּהוּ/tôhûw, to'-hoo, and originates in an unused root word meaning "a desolation (of surface), i.e., desert; figuratively, a worthless thing; adverbially, in vain, confusion, empty place, without form, nothing, (thing of) nought, vain, vanity, waste, wilderness."

The Hebrew word for void, בֹּהוּ/vôhûw, vo'-hoo, means "to be empty; a vacuity, i.e., an indistinguishable ruin; emptiness."

The phrase ṭō·hū va vō·hū is referenced in the book of Jeremiah where these words refer to divine judgment (Jer 4:23). Genesis 1:1, 2 says "and darkness was on the face of the deep. And the Spirit of God was hovering over the face of the waters." The Hebrew word the deep, תְּהוֹם/tehom, defined as "deep, sea, abyss," may also describe "an inhospitable ice planet."

Biblical teachings impart that death did not exist until Adam's fall. Neither death nor dinosaurs are noted in the time period between verse 1 and verse 2. Yet, there is a gap because of the fall of Helel. The fall of this Angel of Light accounts for the chaos as described in verse 2. We do not know how long the earth was *formless and void*. We do not know the age of the earth before God created life on this planet during the six days of creation and rested on the seventh (Exod 20:11). We do know, however, that life-form is approximately six thousand years old according to biblical genealogical accounts as we prepare to enter the seventh millennial day. We must remember, the law of sin and death, entropy, and curse, occurred only after Adam and Eve fell. Hence, the fossil record follows the fall of man.

God desires that humanity be fruitful and multiply; filling the earth with godly offspring in accordance with his divine order. Therefore, God positioned planet Earth strategically to ensure the advancement of life.

Our Great Creator has our best interest at heart. After Adam's fall, however, this planet became unholy ground. God does not tempt, but he does test (Ps 11:5 and Jas 1:13). Life is a series of choices. Will we come into agreement with God our Maker? God gives us a choice. Living on Earth as a disciple is "bootcamp training" for eternity. Discipleship training is learning to walk in the lifestyle of Yeshua.

God is light, and in him is no darkness nor death at all. In order to know the difference between light and darkness there must be a contrast. God created us with the right to choose. God gives everyone the opportunity to choose light or darkness, life or death, love or hate, good or evil.

God created the tree of the knowledge of good and evil to allow for choice. God is love and love allows us to choose. God is light and light allows us to choose darkness. God is good and good allows us to choose evil. Adam and Eve knew no "evil" until they ate from the tree of the knowledge of good and evil. The word *knowledge* has its origin in the word *knowing*. After partaking of the tree of the knowledge of good and evil "their eyes were opened" to the contrast of good and evil. In the process of time, good and evil would be known and experienced by all their offspring.

Evil permeates the nature of fallen man. An evil nature is the absence of God (read Matt 7:11). For example, cold is the absence of heat. Similarly, dark is the absence of light. Evil is the absence of good, the absence of God.

If God did not allow evil to exist, worship and service to him would be governed by obligation—not as a choice through our will.

The book of First John sheds light, love, and life on the nature and reality of knowing God: "This is the message which we have heard from Him and declare to you, that God is light and in Him is no darkness at all. If we say that we have fellowship with Him, and walk in darkness, we lie and do not practice the truth. But if we walk in the light as He is in the light, we have fellowship with one another, and the blood of Jesus Christ His Son cleanses us from all sin" (1 John 1:5–7).

> Beloved, let us love one another, for love is of God; and everyone who loves is born of God and knows God. He who does not love does not know God, for God is love. In this the love of God was manifested toward us, that God has sent His only begotten Son into the world, that we might live through Him. In this is love, not that we loved God, but that He loved us and sent His Son to be the propitiation for our sins. Beloved, if God so loved us, we also ought to love one another. (1 John 4:7–11)

The original state of evil was inactive until Adam disobeyed God. Evil is activated through a compromised moral position due to doubting God's nature and character. God prohibited Adam and Eve from eating from the tree of the knowledge of good and evil. To do so would result in activating the law of sin and death.

The flu virus bears this example. When one is exposed to a flu virus, the virus is activated through a *compromised* immune system. If Adam and Eve had avoided the tree of the knowledge of good and evil—the fruit of which God told them not to eat—evil would not have been activated. *This same tree Satan used as a ploy to deceive them into disobeying God could have been used by Adam and Eve to prove their loyalty to God, thus avoiding sin and death.* Evil comes as a result of doubting the Lord's goodness and truthfulness. The lies of the Dragon (as symbolized in the constellation Draco) lead to sin, curse, and death.

We must know that the Almighty is not the tempter: "Let no one say when he is tempted, 'I am tempted by God'; for God cannot be tempted by evil, nor does He Himself tempt anyone. But each one is tempted when he is drawn away by his own desires and enticed. Then, when desire has conceived, it gives birth to sin; and sin, when it is full-grown, brings forth death" (Jas 1:13–15).

From Eve's perspective the knowledge of good and evil was *pleasant to the eyes, and a tree desirable to make one wise.* The Bible does not imply that the fruit of the tree from which Eve and Adam ate was poisonous, bitter, or unpalatable. Indeed, it may have been flavorful (Gen 3:6).

Adam and Eve's sin against God resulted in "transferred authority." Hence, the poisonous seed of the serpent permeated the nature of mankind. Man was originally created in the image and likeness of God. But man's image was marred in what is known as "the fall of man."

Thank God for giving his people new birth in the Messiah—a new spiritual nature likened to our heavenly Father. When we are born from above, born of the Spirit of God, our spirit is born again. Our mind, will, and emotions must be renewed daily by the word of God. Only then are we able to think, talk, act, eat, and walk after the lifestyle of Yeshua.

When we're "born again" in the Messiah, we receive a brand-new identity. This new identity needs to be cultivated. God's "test of loyalty" is determined on a daily basis as we make decisions in keeping with our love for God and his commandments.

"He who says he abides in Him ought himself also to walk just as He walked" (1 John 2:6).

We must be careful when people say: "Has God really said?" This is what the serpent suggested to Eve: "Now the serpent was more cunning than any beast of the field which the Lord God had made. And he said to the woman, 'Has God indeed said . . .'" (Gen 3:1).

Disobedience to God opened the door to death. Even today, for those in the Messiah (who are kept by the power of God), the enemy cannot burst into our homes and start disrupting our lives with the curse. No, the devil must be resisted and not be allowed to gain a foothold.

Did you know the Hebrew word, שָׁמַע sh'ma "to hear, listen" also means "to obey." "Hear [שָׁמַע/sh'ma], O Israel: The LORD our God is one LORD" (Deut 6:4). Thank God for his instructions regarding "faith": "So then faith comes by hearing, and hearing by the word of God" (Rom 10:17). According to the apostle James, to truly hear is to truly do (Jas 1:22–25).

Prior to Adam and Eve believing the lies of the enemy, they experienced only the goodness of God. His grace and mercy would not be realized, however, until after the fall. God's goodness would be expressed to them in a way they had yet to realize. God graciously clothed Adam and Eve in the covenantal skins of a lamb. These lamb skins were a "foretelling" of Jesus, Yeshua, the Lamb of God, who takes away the sins of the world (John 1:29). Note: Yeshua's substitutionary sacrifice on the cross is portrayed in the constellation Aries.

Adam and Eve's "fall" did not take God by surprise. God foreknew they would fail the test. Yet, his goodness would be expressed to them just as it is to us today. Indeed, God made provision for them and their descendants before the foundation of the world. Prior to Adam and Eve witnessing a withering leaf die and drift to the ground, the constellation Aries, the Ram (a pictorial symbol of Yeshua, the Lamb of God), hovered over the earth as a testimony of God's grace.

When we call upon the name of the Lord, God responds to us according to his great mercies and compassion. In effect, good overcomes evil. *God's love is greater than Satan's hate.* We must know that the Almighty has a 0 percent failure rate. Love never fails. As the classic hymn says: "*Redeeming grace to Adam's race—The saints' and angels' song.*"[1]

---

1. Lehman, "Love of God."

God has allowed evil and the curse to run its course. The only remedy for sin and curse is the gospel. God is giving everyone the right to choose. Thus, the name of God is vindicated for time and eternity: "The LORD is righteous in all His ways, and kind in all His deeds" (Ps 145:17 NASB).

The author of love offers everyone the right to love him who first loved us. People can resist God's grace as did Lucifer. God did not create Satan. God created Lucifer who *became* (chose to become) Satan, meaning שָׂטָן/*sâṭân*, "an opponent; the arch-enemy of good; the adversary."

The love of God afforded Lucifer the opportunity to rebel. And the Lord God offers us the same choice—to make our decision for him. No one is coerced to serve and worship the Lord. A good example of this is found in the prophecy of Isa 7:14–15 concerning the incarnation of the young Messiah himself who would learn to refuse evil and choose good. Messiah passed "the test" and became the *tested, chief cornerstone*, the foundation of the church, the spiritual Israel of God (Isa 28:16; 1 Pet 2:6).

Man was created in the image of God. Mankind is vested with the capacity to choose good or evil (Gen 1:26, 27). God's gift to his original creation was the gift of choice, having the ability to choose. Today, we have the capacity to choose the gospel of the grace of God—and to keep his commandments, which are *holy, just, and good* (Rom 7:12).

Violating God's commandments constitutes sin and places one under the law, meaning "under the penalty of the law (Torah)," where evil consequences are activated.

A tangible example of evil is poison ivy. God created poison ivy as an object lesson. He did not say to eat poison ivy. In fact, we do well not to touch it. The same principle is true when it comes to evil. In the beginning God created the tree of the knowledge of good and evil and said, "*Do not eat from this tree.*" The same principle applies today. We are to choose the good and refuse the evil. We are told not to choose evil, nor cultivate or practice evil. Helel is the author of cultivating evil and he tries to further his cause (through the seed of the serpent) within the human race. Even though God created evil (in the sense of its allowance), he never cultivates evil. God only cultivates good. He is the author of all that is good and wants us to know and experience the same. The Almighty would have us learn the art of knowing and doing good, producing good fruit which endures to eternal life.

"For the LORD is good; His mercy is everlasting, and His truth endures to all generations" (Ps 100:5).

Regarding the original messianic prophecy (Gen 3:15) and the gospel as communicated in the stars, the apostle Paul says: "But I want you to be wise in what is good, and simple concerning evil. And the God of peace will crush Satan under your feet shortly" (Rom 16:19, 20).

The Greek word here for "simple" is ἀκέραιος, ον/*akeraios*, meaning "unmixed with bad; pure, blameless and innocent." We find this same word in the Greek translation of the book of Matthew: "Behold, I send you out as sheep in the midst of wolves. Therefore, be wise as serpents and harmless [*akeraios*] as doves" (Matt 10:16).

The principle of *akeraios* "purity" is also relevant regarding Bible doctrine. We must guard the garden of our hearts. We can pull out the wild weeds and cultivate the pure and good, healthy herds. We can prevent a mixture of the commandments and doctrines of God mingled with the commandments and doctrines of men.

Ultimately, God takes responsibility for evil. This is why he took our sin upon himself on the cross. In fact, the Scripture says, "Having made peace [*shalom*] by the blood of His cross" (Col 1:20).

God will sentence the devil to the lake of fire along with those who reject his offer of salvation. The proverb says: "The Lord has made all for Himself, yes, even the wicked for the day of doom" (Prov 16:4).

God the Creator gives us the right to choose life today, even as he did with Israel of old, and encourages us in what to choose: "I call heaven and earth as witnesses today against you, that I have set before you life and death, blessing and cursing; therefore, choose life, that both you and your descendants may live" (Deut 30:19).

The heavenly constellations display Satan in his various forms: Draco (the Dragon), Scorpio (the Scorpion); Hydra (the Fleeing Serpent); Cetus (the Sea Monster); Serpens (the Serpent Thief). Each receives a fatal death blow to the head. This fatal death blow to the head of these serpentine constellations became a reality when the Messiah thrust upward on the cross to crush this enemy two thousand years ago.

Draco (the dragon's head) is crushed by Hercules (Gibbôr). Scorpio (the scorpion's head) is crushed by the foot of Ophiuchus (the Serpent Restrainer, representing the Messiah). Hydra (the Fleeing Serpent) is crushed by Regal, the foot of the Lion of the tribe of Judah. Cetus (the Sea Monster), has the bands of Pisces (the Fishes) broken off his neck by Aries (the Ram) depicting that one's bond with sin and Satan is forever

broken. Serpens' (the Serpent Thief) head is struck with the club in Gibbôr's right-hand.

Thank God for his faithful promise in providing the *Seed* of the
Woman (Yeshua) to crush the head of the old dragon and give us power
to trample (*darak*) him under foot. Remember, before the first messianic
prophecy was given by the Almighty in Gen 3:15 the Lord already set the
constellations in place, including Hercules (Gibbôr) crushing the head
of Draco.

In summary, Lucifer became Satan. And this "seed of the serpent"—
who morphed into an enormous dragon through devouring the "dust"
of humanity—is passed down through the children of men manifesting
itself in selfishness and sin.

As stated at the beginning of this chapter, the biblical Hebrew name
for Draco is *leviathan, the crooked serpent.* The Hebrew word *Aqallathon*,
or "crooked," originates in the word עָקַל/*aqal*—"to bend, twist, pervert,
crooked justice." An example of the word עָקַל/*aqal* is found in the book of
Habakkuk. "Therefore the law is powerless, and justice never goes forth.
For the wicked surround the righteous; therefore perverse [עָקַל/*aqal*]
judgment proceeds" (Hab 1:4).

Every day in the news we see the manifestation of the seed of the
serpent expressed where perverse (עָקַל/*aqal*) judgment proceeds. The
schemes of the enemy are working behind the scenes. The only way out
of this perversion is to receive the Seed of God, Jesus Christ (the Seed
of the Woman) into one's life by means of the new birth. Through "new
birth," being born from above, born of the Spirit of God, we can join God
in eliminating evil from our lives and the lives of others (see 1 John 3:9).

The key to understanding how God destroyed the enemy is revealed
in an obscure scripture: "But we speak the wisdom of God in a mystery,
the hidden wisdom which God ordained before the ages for our glory,
which none of the rulers of this age knew; for had they known, they
would not have crucified the Lord of glory" (1 Cor 2:7, 8).

Even though Satan was clever in his dealings with God and man,
he did not know how *his* headship (authority) would be crushed according to the ancient prophecy. The arch enemy of God evidently assumed
when the Son of God died on the cross and was resurrected, his (Satan's)
headship authority over Adam's race would remain. How could he know
his headship would be crushed by means of transferred authority? Satan
did not know the Messiah (the last Adam), through his death, burial, and

resurrection would—once, for all—recover the authority Adam and Eve once enjoyed before the fall. Moreover, the Jewish rulers, and the rulers of the Roman governor, were ignorant of the gospel as revealed in the Hebrew scriptures (read: John 12:31, 14:30, 16:11; Acts 3:17, 4:8, 4:26).

The devil tried to coerce Yeshua to bow before his (the devil's) authority: "Then the devil, taking Him up on a high mountain, showed Him all the kingdoms of the world in a moment of time. And the devil said to Him, 'All this authority I will give You, and their glory; for this has been delivered to me, and I give it to whomever I wish. Therefore, if You will worship before me, all will be Yours'" (Luke 4:5–7).

Yeshua did not contest the devil's "authority." He simply rebuked him; directing him to the final authority of God's word. The Greek word for "delivered" in this verse is *paradidómi*, "to hand over, to give or deliver over, to betray." This is what occurred when Adam and Eve sinned in the garden of Eden. There was a transfer of authority—from Adam and Eve to the devil. This authority was, in essence, handed over to the enemy by our first parents, Adam and Eve. In order to recover this authority, the sinless Son of God took it back from the enemy by means of his death, burial, and resurrection.

This concept of transferred authority is illustrated by Yeshua himself in the Gospel of Luke: "When a strong man, fully armed, guards his own palace, his goods are in peace. But when one stronger than he comes upon him and overcomes him, he takes from him all his armor in which he trusted, and divides his spoils. He who is not with Me is against Me, and he who does not gather with Me scatters" (Luke 11:21–23).

Yeshua is the stronger man in this illustration who overtakes the strong man, "Satan," and strips him of his armor and possessions in which he trusted. Hercules/Gibbôr is one of the constellations which represents Yeshua as the stronger man.

The authority of God and his Messiah has always been supreme over all creation, and God's authority has never been successfully contested by the enemy. Only God could intervene to restore the delegated authority originally given to Adam and Eve.

Epic Discovery

During the spring feasts of the Lord, God arranges the sky to "showcase" the constellation Hercules crushing the head of Draco during those hours in which the crucifixion occurred. Could this celestial scene tell the story of the first messianic prophecy (Gen 3:15) given by the Lord in the Bible?

On Wednesday, April 25, AD 31, from 2:00 to 3:00 p.m.—the exact hour of the Messiah's death—the head of Draco, the Dragon (representing Satan), descended and disappeared below Earth's northwest horizon, symbolically being crushed by Hercules, the Mighty Gibbôr. Scripture verifies darkness covering the earth from 12:00 to 3:00 p.m. When Christ hung on the cross, the stars and constellations were possibly visible during those three hours of darkness. This phenomenon would reflect and point to the fulfillment of their heavenly message on Earth—in real-time.

This "pictorial prophecy" illustrates the fatal blow to the authority of the serpent/dragon by the Messiah. This mystery, kept hidden in God, is much like the constellations—hidden by the light of the Sun. Particularly noteworthy, at 12:00 noon when darkness began to cover the earth, Leo, the Lion (representing Yeshua, the Lion of the tribe of Judah) rose in the east while Draco (representing Satan) descended in the northwest.

The heavens are, in effect, announcing the headship authority of the enemy has been removed from the people of God as the Mighty Gibbôr crushes the head of the dragon.

Three days later, on Sunday morning, April 29, AD 31, at 4:30 a.m. the constellation Aries (the Ram) is rising in the east just before sunrise with the Sun aglow at his feet. Simultaneously, Perseus is rising from the dark domain (symbolically) holding Satan's head in his hand.

These two events (the crucifixion and resurrection scene) in astronomy are connected. At the crucifixion, the head of the enemy is crushed, and at the resurrection, the head of the enemy is carried away. This fulfills both the Gen 3:15 and 1 Sam 17:54 prophecies concerning the Messiah's triumph over the enemy of our souls.

The devil's destiny, along with his followers, is death as prophesied in the book of Nahum: "*affliction will not rise up a second time*" (Nah 1:9).

God's cosmic clock is precise to the second and on display in the heavens, allowing us to know and keep the Lord's appointed feast days.

Then God said, "Let there be lights in the firmament of the heavens to divide the day from the night; and let them be for signs and seasons, and for days and years" (Gen 1:14).

According to the first chapter of Genesis, one of the primary purposes in observing the starry universe is to understand its signs and seasons. The Hebrew word sign is אות/*oth*, meaning "a signal, a beacon." The Hebrew word for seasons is מֹועֵד/*moed*, meaning "an appointment, a fixed time, a festival."

The word used here for seasons, מֹועֵד/*moed*, does not mean the four seasons, but rather the "seasons" of the Lord's feast days. The Lord's biblical feast days include the weekly Shabbat (Sabbath), and the annual feast days: Passover, Unleavened Bread, Firstfruits, Pentecost, the Feast of Trumpets, Day of Atonement, and the Feast of Tabernacles. The lunar calendar designates the times and seasons in which to honor the Lord's appointed feast days. God's cosmic clock is precise to the second.

Yeshua instructs us to be attentive to the signs in the heavens. In doing so, we can watch for the cosmic signs of his second coming and point others to the gospel. "And there will be signs in the sun, in the moon, and in the stars" (Luke 21:25).

As we turn the cosmic clock back to the crucifixion and resurrection of the Messiah, we are viewing the Creator's אות/*oth*, "a signal, a beacon" on the Lord's מֹועֵד/*moed*, "feast days," specifically Passover, Unleavened Bread, and Firstfruits of AD 31.

Stellarium image (close-up): This astronomical configuration occurred (Nisan 14) April 25, AD 31, (as seen from Jerusalem) during Passover at the crucifixion of Messiah. At 12:00 noon when darkness began covering the earth, Leo, the Lion (representing Yeshua, the Lion of the tribe of Judah) rose in the east while Draco (representing Satan) descended in the northwest. Draco's head represents Satan's authority being crushed beneath the feet of the Great Gibbôr, Messiah Yeshua.

Stellarium image (scale size): This astronomical configuration occurred (Nisan 14) April 25, AD 31 (as seen from Jerusalem) during Passover at the crucifixion of Messiah. At 12:00 noon when darkness began to cover the earth, Leo, the Lion (representing Yeshua, the Lion of the tribe of Judah) rose in the east while Draco (representing Satan) descended in the northwest.

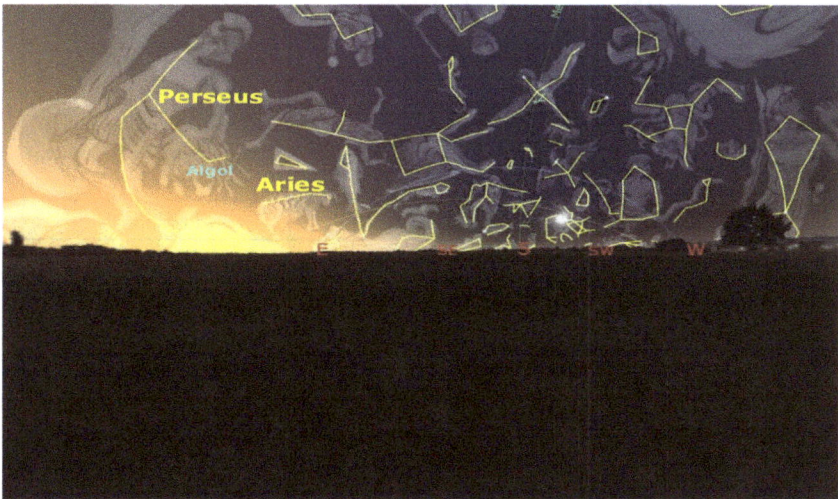

Stellarium image: This astronomical configuration occurred on Sunday morning, April 29, AD 31, at 4:45 a.m. The constellations Aries and Perseus rise at 4:30 a.m. just before sunrise depicting Satan's head (Al Gol) in the left hand of Perseus. Perseus in Hebrew means "breaker" and the star *Al Gol* is an apparent abbreviation of (Gulgoleth) the Hebrew word for "skull, head." This imagery indicates the headship authority of Satan has been utterly broken and removed from the people of God. There are several other significant signs occurring here. Yeshua has indeed conquered Satan and death!

Constellation Name	Translation
Draco (Latin)	Dragon
Livyathan Aqallathon Nachash (Hebrew) (Isa 27:1)	leviathan the crooked serpent
At-Tinneen (Arabic)	the dragon
Drakôn (Greek)	the dragon
tiān lóng zuò (Chinese)	the heaven dragon constellation

Enclosed are derivatives and similar words in ancient languages which help clarify the core meaning of this constellation/star name.

Ancient Hebrew/ Aramaic	Phonetic Spelling	Summarized Meaning
Hebrew לִוְיָתָן	Livyathan	"serpent," a sea monster or dragon (Isa 27:1) from: לָוָה/lâvâh, to twine, i.e., (by implication) to unite, to join
Aramaic	Liviatan	Leviathan
Hebrew עֲקַלָּתוֹן	Aqallathon	crooked, twisted the constellation Draco;— לִוְיָתָן נָחָשׁ ע (Isa 27:1) from עָקַל/aqal to bend, twist wrong, pervert, crooked justice
Hebrew נָחָשׁ	Nachash	serpent (Isa 27:1)
Hebrew דָּרַךְ	darak	tread, march
Aramaic ܕܪܟܬܐ	drkta	treading out
Aramaic ܕܪܟܐ	drk	to tread; to achieve control of

Ancient Hebrew/ Aramaic	Phonetic Spelling	Summarized Meaning
Aramaic	qllh, qllt	curse
Aramaic	qltn	one who has lost a testicle

Ancient Gentile Language	Phonetic Spelling	Summarized Meaning
Greek Ὄφιν Φεύγοντα	Ophin pheugonta	flying serpent (the Septuagint—Isa 27:1)
Arabic	lawwā	to bend, to coil, to twist
Persian	nawashtan	to fold, twist, coil
Chinese	zhěn	crooked, obstinate, twist a cord
Hindi	veshthan	coil
Amharic/Ethiopian	leewatan	leviathan
Tigrinya/Ethiopian	liwyatan	leviathan
Latin	Dracō	dragon
Greek Δράκων	drakōn	dragon
Assyrian	da: ik	to trample, to tread heavily

Scripture References

Isa 27:1; Gen 3:15; Ps 91:13; Isa 14:12–17; Ezek 18:12–19; Isa 45:6–7; Jas 1:13–15; Gen 3:6; 1 John 2:6; Gen 3:1; Ps 145:17; Isa 7:14–15; Gen 1:26–27; 1 Tim 2:13–14; Rom 7:12; Ps 100:5; Rom 16:19–20; Matt 10:16, 24:4–5; Col 1:20; Prov 16:4; Deut 30:19; Isa 26:19–21; 1 Cor 15:51–55; Hab 1:4; Ps 125:5; 1 Cor 2:7–8; Luke 4:5–7, 11:21–23; Nah 1:9; John 12:31, 14:30, 16:11; Acts 3:17, 4:8, 4:26

The Stars

Thuban

THE STAR NAME THUBAN, an Arabic word for "snake," occupied the "polestar" position (the central position of the celestial sphere) from approximately 3700 to 1900 BC.

Thuban became more centralized as the pole star in 2830 BC and would discontinue as the polestar in approximately 1900 BC. The central position of Thuban corresponds with the scriptural designation of Satan as the god and prince of this world (2 Cor 4:3–4; John 14:30).

Polaris in Ursa Minor (Dover, the Lesser Sheepfold) became the polestar in approximately AD 1100. This constellation represents the inheritance of the saints in light (Col 1:12–14). Ursa Minor (Dover, the Lesser Sheepfold, "little flock") is a picture of the Messiah's kingdom administration. This constellation reminds us that the Messiah has deposed the enemy from his throne and restored the authority originally given to Adam and Eve to his people.

Draco is wrapped around Ursa Minor (Dover, the Lesser Sheepfold). This symbolically shows the enemy is surrounding the Lord's heavenly sheep but is incapable of touching God's elect. In keeping with this cosmic scene God has given us such promises as:

> And I give them eternal life, and they shall never perish; neither shall anyone snatch them out of My hand. (John 10:28)

> No evil shall befall you, nor shall any plague come near your dwelling. (Ps 91:10)

> Do not touch My anointed ones, and do My prophets no harm. (Ps 105:15)

> We know that whoever is born of God does not sin; but he who
> has been born of God keeps himself, and the wicked one does
> not touch him. (1 John 5:18)

Polaris became more centralized around 1967, the year Jerusalem was restored as the capital of Israel.

Polaris will become even more centralized as the hub of the celestial heavens in AD 2100. From AD 2700 onward, Polaris gradually loses its position as polestar. Due to the precession of the equinoxes, the star Errai, or "the Shepherd," in Cepheus (representing Messiah, the King) will become the northern polestar in approximately AD 3000 and will make its closest approach around AD 4000. This also agrees with the timeline of the millennial reign of the Messiah beginning in the near future.

Following the millennial reign of Messiah, creation will enter the eighth millennium day in which our faithful Creator makes all things new. This newness is exemplified in the star Al Rai, emblematic of "our great Shepherd, King of the universe" as it becomes the hub of the celestial heavens. At this time, creation will be free from entropy and reside in universal righteous perfection, forevermore. Let us look up and anticipate the glorious future, we have who put our trust in Yᵉhôvâh's Messiah, Yeshua!

Polaris represents Messiah's central authority as experienced by the people of God. This authority will continue through the millennial reign of Messiah. When Thuban was replaced by Polaris, it confirmed to the universe that the transfer of authority was accomplished. As we allow the Messiah and his word to be the center of our universe, a greater degree of balance is experienced in our lives. We become finely tuned by keeping our focus on the Messiah and his word as the hub of our lives.

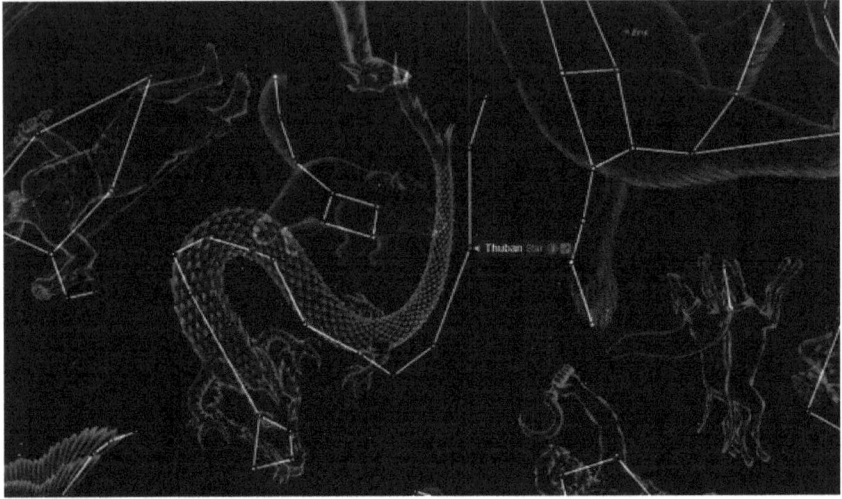

Starry Night image: Thuban in Draco was the polestar from 3700 to 1900 BC and became centralized around 2830 BC (image courtesy of Starry Night Astronomy, 2023, all rights reserved).

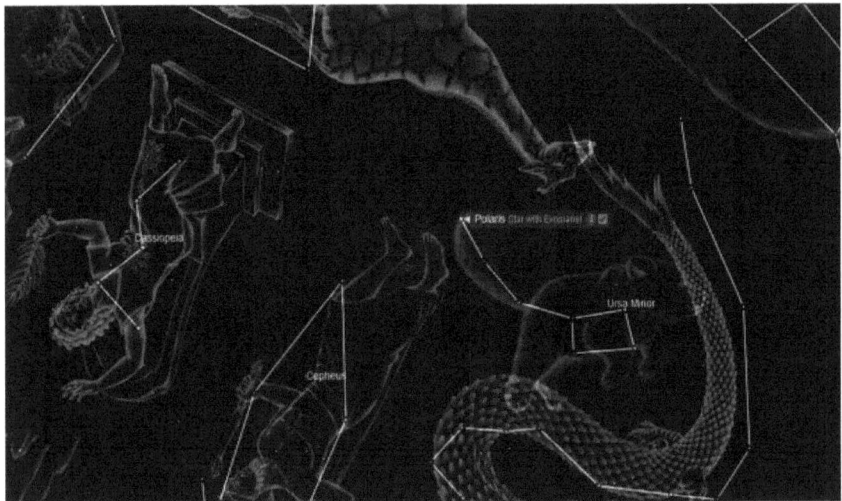

Starry Night image: Polaris, in Ursa Minor (Dover, the Lesser Sheepfold), has been the polestar since AD 1100. Polaris became more centralized around 1967, the year Jerusalem was restored as the capital of Israel. In AD 2100 Polaris will have a greater degree of balance as the hub of the celestial heavens (image courtesy of Starry Night Astronomy, 2023, all rights reserved).

Conventional Star Name/Meaning	Name/Meaning Confirmed or Corrected
Thuban—the subtle	confirmed: snake (inferred: the subtle)

Enclosed are derivatives and similar words in ancient languages which help clarify the core meaning of this constellation/star name.

Ancient Gentile Language	Phonetic Spelling	Summarized Meaning
Arabic	*ṭu'bān*	snake
Tamil	*taṭṭāṉ*	snake

Scripture References

2 Cor 4:3–4; John 14:30; Gen 3:1; John 10:28; Pss 91:10, 105:15; 1 John 5:18

Ras Taban and Al Waid

This star has two names: Ras Taban and Al Waid constitute one of the four stars in the head of Draco. The star name Ras Taban means "head of the snake" in Arabic. Ras Taban is a combination of two Arabic words: *ras*, meaning "head," and *Taban*—likely a lost pronunciation of *Thuban*, meaning "snake."

God, in the first ancient messianic prophecy, mentions the head of the serpent: "And I will put enmity between you and the woman, and between your seed and her Seed; He shall crush your head, and you shall bruise His heel" (Gen 3:15).

When the Messiah died on the cross to save us all from Satan's power, he crushed the Dragon's head. In the process of crushing the Dragon's head the Messiah's heel was bruised. The crushing of the head of the serpent/dragon by the Messiah was a mortal blow to the serpentine powers. The first blow completely crushed Satan's authority. In the final analysis, the Lord will crush Satan's existence.

"Inasmuch then as the children have partaken of flesh and blood, He Himself likewise shared in the same, that through death He might destroy him who had the power of death, that is, the devil, and release those who through fear of death were all their lifetime subject to bondage" (Heb 2:14, 15).

When Adam and Eve listened to the lies of the enemy serpent, they lost their dominion. They handed their authority over to the devil. The only way to recover this authority would be for the sinless Son of God to—legally—take it back by means of his death, burial, and resurrection—completing the "transfer" of authority. It was necessary for the Messiah to not only die for our sins on the cross, but to rise from the dead on the third day for our justification: "who was delivered up because of our offenses, and was raised because of our justification" (Rom 4:25).

The restoration of dominion is depicted in the constellation Corona. The dominion Satan stole from Adam and Eve is restored in Yeshua, the last Adam. In the Messiah we are legally free from the headship authority of the enemy, just as the Messiah is free and lives forevermore. Messiah is our head. It is he who has reconciled us to the Father of life.

God's eternal purpose is revealed through the cross. Through the cross, God put to death the enmity expressed between Jew and Gentile passed down through the seed of the serpent. In Messiah, God is creating out of two (Jew and Gentile) *one new man*, so making peace. The Hebrew word for peace is *shalom*, meaning "nothing missing, nothing lacking, complete well-being (in the Messiah, the Prince of Peace.)" Through the Messiah we have peace *with* God and the peace *of* God that passes all understanding.

> For He Himself is our peace, who has made both one, and has broken down the middle wall of separation, having abolished in His flesh the enmity, that is, the law of commandments contained in ordinances, so as to create in Himself one new man from the two, thus making peace, and that He might reconcile them both to God in one body through the cross, thereby putting to death the enmity. And He came and preached peace to you who were afar off and to those who were near. For through Him we both have access by one Spirit to the Father. (Eph 2:14–18)

After rising from the dead, the first words Yeshua spoke to his disciples were: שָׁלוֹם עֲלֵיכֶם/*shālôm 'alêichem*, which means "peace be upon you" (John 20:19).

Yeshua saves us individually and wants us to live together in his shalom.

The star name Al Waid appears to be based on the Hebrew אָבַד/ *avad*, meaning "to wander away, lose oneself; to perish, destroy."

An example of this word in the Hebrew Bible is as follows: "For the LORD knows the way of the righteous: but the way of the ungodly shall perish [*avad*]" (Ps 1:6).

The Salkinson-Ginsburg Hebrew New Testament translation of John 3:16 uses this Hebrew word (*avad*): "For God so loved the world that He gave His only begotten Son, that whoever believes in Him should not perish [*avad*] but have everlasting life" (John 3:16).

We must realize that hell is prepared for the devil and his angels as exhibited by the star name Al Waid. Hell is not intended for mankind. When one is sentenced to hell by the Almighty, it is because that one has rejected God's offer of forgiveness (through the gospel) and has sided with the enemy and therefore receives the same punishment, ultimately perishing (*avad*) in the lake of fire.

"Then He will also say to those on the left hand, 'Depart from Me, you cursed, into the everlasting fire prepared for the devil and his angels'" (Matt 25:41).

Who will you blame should you perish in eternal flames? Listen friend, there's no other way. God has provided the way of escape. Yeshua is the only way of salvation. Yeshua alone is able to save us from death, hell, and the grave. Yeshua alone is able to give us eternal life.

I implore you, be saved today while it is called today!

Conventional Star Name/Meaning	Name/Meaning Confirmed or Corrected
Ras Taban—the head of the subtle	confirmed: head of the snake inferred: the head of the subtle

Enclosed are derivatives and similar words in ancient languages which help clarify the core meaning of this constellation/star name.

Ancient Hebrew/ Aramaic Ras Taban	Phonetic Spelling	Summarized Meanings
Hebrew ראש	rosh	head

Ancient Gentile Language Ras Taban	Phonetic Spelling	Summarized Meanings
Arabic	ras	head
Arabic	ṭu'bān	snake
Persian	tābān	light, luminous, resplendent, radiant, dazzling, brilliant, shining, glittering

Conventional Star Name/Meaning	Name/Meaning Confirmed or Corrected
Al Waid—who is to be destroyed	confirmed

Enclosed are derivatives and similar words in ancient languages which help clarify the core meaning of this constellation/star name.

Ancient Hebrew/ Aramaic Al Waid	Phonetic Spelling	Summarized Meanings
Hebrew אָבַד	avad	to wander away, lose oneself; to perish, destroy
Aramaic ܐܒܕ	abd	perish

Ancient Gentile Language Al Waid	Phonetic Spelling	Summarized Meanings
Arabic	qāḍa	destroy
Hindi	vadh	destroy

Scripture References

Ras Taban: Gen 3:15; Heb 2:14–15; Eph 2:14–18; John 20:19
Al Waid: Ps 1:6; John 3:16; Matt 25:41

El Tanin

The star name El Tanin is the Arabic word for "the snake." The word *tanin* is a derivative of the Hebrew תַּנִּין/*tannin*, meaning "serpent, dragon, sea monster." An example of the word *tannin* is found in the book of Exodus: "For every man threw down his rod, and they became serpents [*tannin*]. But Aaron's rod swallowed up their rods" (Exod 7:12).

This symbolizes God's power over the power of the enemy. Aaron's rod swallowed up the rods of the wise men, sorcerers, and magicians of Egypt. In like manner, the Messiah swallowed up the curse on the cross. When we apply the cross to our lives, God's miracle-working power swallows the curse.

The name *tannin* is found in the Salkinson-Ginsburg Hebrew New Testament translation of the Greek text of Revelation: "He laid hold of the [תַּנִּין/*tannin*] dragon, that serpent of old, who is the Devil and Satan, and bound him for a thousand years" (Rev 20:2).

Thank the Lord for delivering us from the rebellious dragon and his serpentine powers.

Conventional Star Name/Meaning	Name/Meaning Confirmed or Corrected
El Tanin—the long serpent	confirmed

Enclosed are derivatives and similar words in ancient languages which help clarify the core meaning of this constellation/star name.

Ancient Hebrew/ Aramaic *El Tanin*	Phonetic Spelling	Summarized Meanings
Hebrew תַּנִּין	tannin	serpent, dragon, sea monster
Hebrew הַתַּנִּין	hatannin	the dragon Rev 20:2 (Heb NT)
Hebrew שָׂטָן	sâṭân	adversary
Aramaic ܐܢܝܢܐ	tnyna	dragon, monster
Aramaic	tnyn, tannīn, tannīnā	sea serpent, constellation dragon

Ancient Gentile Language	Phonetic Spelling	Summarized Meanings
Arabic	tinnīn	dragon
Persian	tinnīn	dragon
Assyrian	tan ' ni: na	dragon
Akkadian	tāntu	the sea, the ocean
Turkish	ju' łan	dragon, serpent
Chinese	téng	flying dragon

Scripture References

Exod 7:12; Rev 20:2

Giansar

Biblical scholar and theologian Ethelbert William Bullinger, better known as E. W. Bullinger, author of *The Witness of the Stars*, defines the Hebrew word *Giansar* as "the punished enemy."[1] The star name Giansar is phonetically similar to the Hebrew word יָסַר/*yâsar*, meaning "to chasten, punish."

The Semitic root consonants (*gnsr*) of Giansar may be derived from two ancient Hebrew words: גַּן/*gan*, meaning "garden," and שַׂר/*sar*, meaning "prince." The two words combined mean "prince of the garden."

---

1. Bullinger, *Witness of the Stars*, 60.

Ezekiel says this regarding Helel who became Satan: "You were in Eden, the garden of God" (Ezek 28:13).

Could this name be reflective of the dominion Satan stole from Adam and Eve in the beginning? When God banished Adam and Eve from the garden of Eden after succumbing to the lies of Satan, Satan essentially became "prince of the garden."There is no evidence that the devil was cast out of the garden while it remained. But after Adam and Eve were banished from the garden he had no reason to remain. Who was left to tempt? By means of transferred authority Satan became the prince and god of the world. But God has always maintained ultimate authority. The Messiah crushed Satan's headship authority at his death, burial, and resurrection, so Satan is no longer god over the Lord's people. The redeemed of the Lord are participants of God's authority in Christ. And just as the star Giansar is positioned at the end of the tail of Draco, Satan's reign over the children of Adam (lost humanity) will come to an end at the end of this age.

Paradise lost is ultimately restored by the Messiah. May we look to Yeshua with glorious expectation! For Yeshua is the true Prince who restores all things and makes all things new.

Conventional Star Name/Meaning	Name/Meaning Confirmed or Corrected
Giansar—the punished enemy	confirmed: also means, "prince of the garden"

Enclosed are derivatives and similar words in ancient languages which help clarify the core meaning of this constellation/star name.

Ancient Hebrew/ Aramaic	Phonetic Spelling	Summarized Meaning
Hebrew יָסַר	*yâsar*	to chasten, punish
Aramaic	*ysr*	to chastise
Aramaic	*AaSaR*	bind
Semitic Roots	*gzr*	to cut off, destroy

Ancient Hebrew/ Aramaic	Phonetic Spelling	Summarized Meaning
Semitic Roots	śrr	common Semitic noun *śarr-, prince, king.
Hebrew גַּ	gan	garden
Hebrew גִּנָּה	ginnâh	garden
Aramaic	G'aNT,oA	garden

Ancient Gentile Language	Phonetic Spelling	Summarized Meaning
Arabic	jazā	punish
Assyrian	i: ' ṣa: ra	to bind, to take prisoner
Assyrian	sar	chief
Ancient Egyptain	ser	official
Hindi	sardaar	prince
Sumerian	sar	king of
Persian	sar	head, top, commander
Coptic	sa r	man
Egyptian	sr	nobleman, magistrate, ram, sheep
Arabic	janna	garden
Egyptian Arabic	gineena	garden
Persian	janān	garden
Assyrian	gan na	garden

Scripture References

2 Chr 10:11; Lev 26:18; Deut 22:18

El Atik

The star name El Atik (in Draco) means "antique, old" in Arabic and Persian. The corresponding Hebrew word to best fit the context of this constellation, עָתַק/âthaq, means "to remove (intransitive or transitive) figuratively, to grow old; specifically, to transcribe, copy out, leave off, become, wax old, remove."

Atik is phonetically similar to our English word antique. An example of this word in the Hebrew Bible appears in Ps 6: "My eye wastes away because of grief; It grows old [עָתֵק/*âthaq*] because of all my enemies" (Ps 6:7).

Atik describes the condition of the enemy who sinned against God. Although we do not know the age of the cherub who became Satan, we know his days are numbered.

"Woe to the inhabitants of the earth and the sea! For the devil has come down to you, having great wrath, because he knows that he has a short time" (Rev 12:12).

The Hebrew word עָתֵק/*âthaq* relates to the constellation Draco and to the star Thuban, in Draco, which was literally removed from its position as the polestar. Yeshua conquered death, hell, and the grave by crushing the dragon's head. This "crushing" resulted in the transfer of authority. The headship authority of the old dragon (the prince of the world) *is* forever removed from the people of God.

"So, the great dragon was cast out, that serpent of old, called the Devil and Satan, who deceives the whole world; he was cast to the earth, and his angels were cast out with him" (Rev 12:9).

Found in the gospel of Luke, an astronomical passage of Scripture helps us maintain the proper perspective regarding the demise of Satan and his serpentine powers. In the Messiah, we can tread upon these powers in our pursuit for God and the salvation of souls.

> Then the seventy returned with joy, saying, "Lord, even the demons are subject to us in Your name." And He [Yeshua] said to them, "I saw Satan fall like lightning from heaven. Behold, I give you the authority to trample on serpents and scorpions, and over all the power of the enemy, and nothing shall by any means hurt you. Nevertheless, do not rejoice in this, that the spirits are subject to you, but rather rejoice because your names are written in heaven." (Luke 10:17–20)

We can rejoice with Yeshua, in Spirit, because the God of the Bible writes the final chapter and he writes our names in the Lamb's Book of Life!

Conventional Star Name/Meaning	Name/Meaning Confirmed or Corrected
El Atik—the fraudulent	corrected: to grow old, the antique, to remove

Enclosed are derivatives and similar words in ancient languages which help clarify the core meaning of this constellation/star name.

Ancient Hebrew/ Aramaic	Phonetic Spelling	Summarized Meanings
Hebrew עָתַק	*athaq*	to remove (intransitive or transitive) figuratively, to grow old; specifically, to transcribe—copy out, leave off, become (wax) old, remove
Hebrew עָתְקוּ	*ā·ṭə·qū*	become old
Aramaic עַתִּיק	*at·tēk*	ancient, advanced, aged, old, taken away
Aramaic עָתַק ܥܰܬܶܩ	*T,eQ*	old

Ancient Gentile Language	Phonetic Spelling	Summarized Meanings
Arabic	*El*	Arabic definite article, meaning "the"
Arabic	*atīq*	old
Arabic	*aṭāqa*	to endure
Assyrian	*an ' ti ka*	antique
Persian	*atīq*	antique
Turkish	*antic*	antique, old
Akkadian	*etāqu*	to cross over, proceed along
Sanskrit	*atikrAntayauvana*	oldish
Chinese	*anti-ku*	antique
Arabic	*atiq*	shoulder

Scripture References

Job 9:5, 21:7, 14:18; Ps 6:7; Rev 12:9–12; Luke 10:17–20

Grumium

The origin of the star name Grumium (pronounced GROO-mi-um) is not readily available. In his book, *The Witness of the Stars*, however, biblical scholar and author, E.W. Bullinger offers the Hebrew definition of *Grumium* as, "the subtle."[2] *Grumium* is phonetically similar to the Hebrew word עָרוּם/*Ârûwm*, meaning "cunning (as in a negative connotation): crafty, prudent, subtle."

Initially, the word עָרוּם/*Ârûwm* is used within the context of the serpent in the garden of Eden: "Now the serpent was more cunning [*Ârûwm*] than any beast of the field which the Lord God had made. And he said to the woman, 'Has God, indeed, said, 'You shall not eat of every tree of the garden?'" (Gen 3:1).

Satan, the enemy of God, is clever. He is the master of deception specializing in taking truth and twisting it through agents of Adam's race. He attempts to make "good sound evil" and "evil look good." In order to be equipped to discern good from evil we must know the word of God for ourselves.

We must realize the devil is the deceiver of this world, and that his greatest deception operates within the realm of religion. For example, God says, "Remember the Sabbath day, to keep it holy. Six days you shall labor and do all your work, but the seventh day is the Sabbath of the Lord, your God" (Exod 20:8–10). People sometimes suggest excluding this commandment, labeling it "out-of-date," or, "Old Testament." Yet, Yeshua says, "he is Lord of the Sabbath (Shabbat)" and "if you love me, keep my commandments" (Mk 2:27-28; John 14:15). As for me and my house, we honor the Lord's Shabbat from Friday evening to Saturday evening. The Shabbat is truly a beautiful weekly gift from the Lord. When we study Scripture in light of his absolutes, we find harmony and continuity from Genesis to Revelation.

Another example of the subtlety of the serpent is found in the apostle Paul's comment on the mystery of lawlessness: "For the mystery of lawlessness is already at work; only He who now restrains *will* do so until

---

2. Bullinger, *Witness of the Stars*, 60.

He is taken out of the way" (2 Thess 2:7). The Greek word, ἀνομία/*anomia* is defined as "lawlessness, illegality, violation of law, transgression of the law, unrighteousness."

The Torah (Hebrew scriptures) is the foundational text of our faith. The Scriptures are very clear regarding the finality of God's law: "Do we then make void the law through faith? Certainly not! On the contrary, we establish the law" (Rom 3:31). Alas, "Torah-lessness" has been taught within the context of Christianity. Therefore, "the 'mystery' of lawlessness" remains a mystery.

Love without law is not love. From Genesis to Revelation, the Bible teaches that we are to love and fulfill God's law (Matt 5:17–20; Rom 8:1–4). The apostle John states, "For this is the love of God, that we keep His commandments. And His commandments are not burdensome" (1 John 5:3). By God's grace, we are free to keep his commandments. And the true test of our love for God is our loyalty to the Lord lived out in everyday life.

Praise the Lord! The fear of the Lord enables us to refrain from evil. The love of the Lord enables us to do what is right. Therefore, let us not entertain the subtlety of the serpent as exemplified through this star name, Grumium. Let us be faithful followers of the Lord of heaven and Earth.

Won't you pray with me:

Lord, please forgive and cleanse me of my sins. Thank you, Yeshua, for dying for me on the cross. I receive you into my heart and life. From this time forth I choose to live for you according to your living word, the Bible. Lead me to be water baptized. Fill me with your precious Holy Spirit. Father, help me prepare for the soon coming of Yeshua, so that I may be numbered among the redeemed, in the name and blood of your Son, my Savior, Yeshua. Amen.

Conventional Star Name/Meaning	Name/Meaning Confirmed or Corrected
Grumium—the subtle	confirmed

Enclosed are derivatives and similar words in ancient languages which help clarify the core meaning of this constellation/star name.

Ancient Hebrew/ Aramaic	Phonetic Spelling	Summarized Meaning
Hebrew עָרוּם	*Ârûwm*	cunning (in a negative sense); crafty, subtle From root word עָרַם/ *'âram* (to be (or make) bare; to be cunning in a negative sense; to be aware
Aramaic	*rammāyū, rammāyūṯā*	deceit

Scripture References

Gen 3:1; Prov 14:15; Rev 20:2; Matt 10:16; 2 Thess 2:10–12

Appendix 1

Passover Crucifixion Date

The Messiah's Crucifixion and Resurrection Date Analysis

Criteria:	point 1	point 2	point 3	point 4	point 5	point 6	Criteria Totals
Year	**14th day of Nisan (Passover Crucifixion)**	Estimated age of Yeshua at his death and resurrection	Hours in the heart of the Earth/tomb	Position of the sun	Passover blood moon partial or full (peak time of eclipse)	Sukkot blood moon partial or full (peak time of eclipse)	
AD 27 - AD 34	Day and Hour Passover full moon as seen from Jerusalem (according to Stellarium and TimeandDate.com). The biblical day begins in the evening at sunset.	Based on the sign of Messiah's birth (Rev 12:1) on Trumpets 9-11-3 BC. Yeshua was about 30 years old when he started his ministry. (Luke 3:23)	The biblical sign of Jonah. Three days and three nights (72 hrs.) (Matt 12:40) (Jonah 1:17) Resurrected on the third day. (Luke 24:46-47)	The sun in Aries, (the Ram) during Passover	Sunlight, refracted by Earth's atmosphere, produces a reddish color on the moon. (A partial lunar eclipse can cause the moon to appear red.) Joel 2:28-32; Acts 2:17-21; Matt 24:29; Luke 21:25		
AD 27	Wed. Apr. 9 @ 7:00 p.m.	28.6	72	Aries	no	no	3
AD 28	Tue. Apr. 27 @ 3:00 p.m.	29.7	96	Aries	no	no	1
AD 29	Sun. Apr. 17 @ 5:00 a.m.	30.7	144	Aries	no	no	1
AD 30	Thu. Apr. 6 @ 10:00 p.m.	31.6	60	Aries	no	no	2
AD 31	Wed. Apr. 25 @ 10:00 p.m.	32.7	72	Aries	10:31 p.m. partial/visible (in Scales)	5:31 a.m. partial/visible 10-19-31	6
AD 32	Mon. Apr. 14 @ 11:00 a.m.	33.7	120	Aries	11:25 a.m. full/not visible (in Scales)	3:06 p.m. full/not visible 10-7-32	2
AD 33	Fri. Apr. 3 @ 6:00 p.m.	34.6	36	Aries	5:06 p.m. partial/not visible (above Scales)	6:08 a.m. partial/visible 9-27-33	2
AD 34	Thu. Apr. 22 @ 10:00 a.m.	35.7	60	Aries	no	no	2

The Passover crucifixion date, Wednesday, April 25 AD 31, coincides with the three days and three nights (seventy-two hours) the Messiah's body lay in the tomb and arose on the third day (Matt 12:40). Joel's prophecy, quoted by the apostle Peter on Shavuot/Pentecost, indicates a blood moon was witnessed within the context of the

Passover crucifixion (Acts 2:19–20; Joel 2:28–32).
Note: When the women arrived at the tomb early Sunday morning (the Feast of
Firstfruits, April 29) the Messiah had already risen!

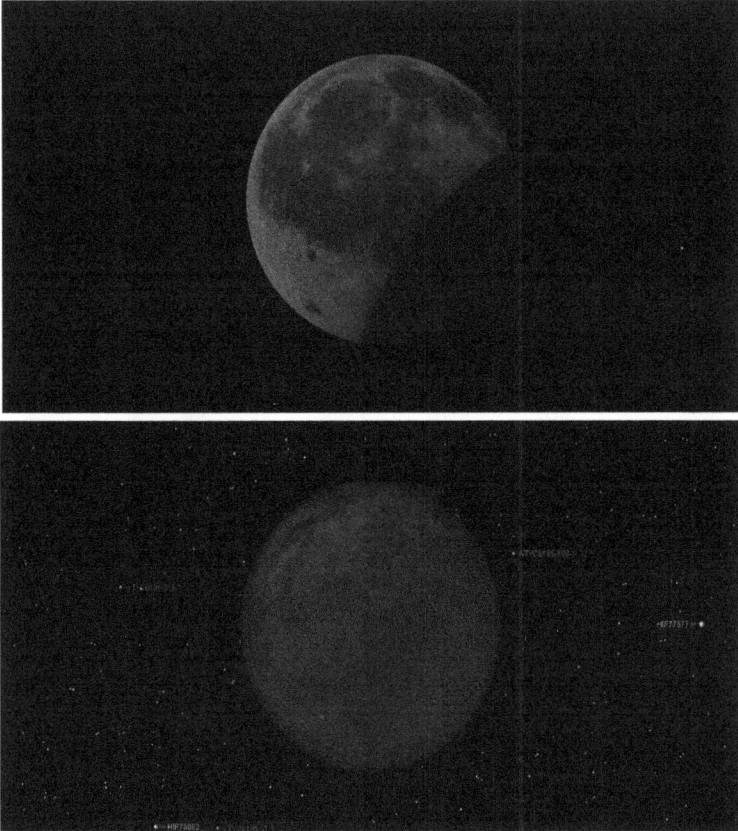

TOP: Stellarium Image. BOTTOM: Starry Night Image (image courtesy of Starry
Night Astronomy, 2023, all rights reserved).
As seen from Jerusalem: On Passover, April 25, AD 31, this partial lunar eclipse
reached totality at 10:31 p.m. About a third of the Moon was totally immersed in
Earth's shadow, casting a reddish hue. Atmospheric dust may have intensified its
redishness. Note: According to Stellarium, Passover blood moons of AD 32 and AD
33 were not visible from Jerusalem.

Appendix 2

Hourly Signs of the Messiah's Death, Burial, and Resurrection

ACCORDING TO SCHOLARLY CONSENSUS, the Messiah was crucified on Passover, (Nisan 14) Wednesday, April 25, AD 31 and rose from the dead, Sunday morning, April 29. I concur.

Using astronomy software, I discovered amazing astronomical signs in the heavens over Jerusalem when reversing the universe to Wednesday, April 25 and Sunday morning, April 29, AD 31.

The Messiah was nailed to the cross at 9:00 a.m. and died at 3:00 p.m. The Bible documents darkness as "covering the land" from 12:00 noon to 3:00 p.m. during Christ's crucifixion. Its possible the stars and constellations were visible during those three hours of darkness as Christ hung on the cross. Regardless, they were present in the "backdrop" of the sky. This celestial scene, portraying and pointing to the fulfillment of their heavenly message on Earth (real-time), is positionally similar during the spring feasts of the Lord over the last two thousand years. Biblical days are reckoned from sunset to sunset. Constellations ascending and descending at the Earth-sky horizons during evening and morning hours on the Lord's feast days appear to be biblically significant. The Messiah fulfilled the spring feasts at his first coming and will fulfill the fall feasts at his second coming.

Signs of His Death: Passover
(Nisan 14/April 25, AD 31)

1. Wednesday, April 25, AD 31, at 12:00 p.m. (the fourth of his seven-hour crucifixion), Leo, the Lion (representing Yeshua, the Lion of the tribe of Judah), ascended in the east, while Draco, the Dragon (representing Satan) descended in the west.

2. Wednesday, April 25, AD 31, at 12:00 noon, Cygnus, the Swan, known as the Northern Cross, descended in the northwest horizon as the Messiah laid down his life for us on the cross.

3. Wednesday, April 25, AD 31, the Pleiades crosses the meridian between 11:50 a.m.–12:00 noon and Orion, the Mighty Warrior crossed the meridian between 1:30–2:15 p.m.

4. Wednesday, April 25, AD 31, from 2:00 p.m. to 3:00 p.m. (the hour of the Messiah's death), the head of Draco, the Dragon (representing Satan, the enemy), descends and disappears below the northwest horizon while symbolically being crushed by Hercules, the mighty Gibbôr (representing the Messiah).

5. Wednesday, April 25, AD 31, Gemini, the Twins cross the meridian in zenith position at 3:00 p.m. during Messiah's final hour on the cross.

6. Wednesday, April 25, AD 31, at 6:00 p.m., the constellation Perseus, the Warrior descended beneath the western horizon (the Sun aglow at his feet), symbolically holding Satan's head (star: Al Gol) in his left hand.

7. During the AD 31 spring feasts of the Lord, Jupiter (defined in Hebrew as "righteousness"), is in Aries, the Ram. Yeshua, is our righteousness: "My righteous Servant shall justify many, For He shall bear their iniquities" (Isa 53:11).

Signs of His Burial: Unleavened Bread/
Holy Convocation/High Day-Sabbath
(Nisan 15/April 25, AD 31)

8. Wednesday, April 25, AD 31, at 7:00 p.m. approximately when the Messiah's tomb was sealed, the image of the constellation Aries

(representing Yeshua, the Passover Lamb) descended beneath the western horizon after sundown.

9. Wednesday, April 25, AD 31, at approximately 7:00 p.m. as the Messiah's tomb was sealed, Leo, the Lion, is in zenith position. Three days and three nights later "on resurrection day" the Lion is in zenith position after sundown.

10. Wednesday, April 25, AD 31, as the Messiah's tomb was sealed, Crux, the Southern Cross, is descending in the southern horizon at 7:00–11:00 p.m.

11. Wednesday, April 25, AD 31, at 7:00 p.m. approximately when the Messiah's tomb was sealed, a full Passover moon rose over Jerusalem after sundown. Two and a half hours later a partial lunar eclipse appeared over Jerusalem. This lunar eclipse (blood moon) peaked at 10:31 p.m. The duration of the eclipse lasted 115 minutes: 9:35 to 11:30 p.m. (1 hour and 55 minutes).

12. April 25, AD 31, thirty minutes after the blood moon lunar eclipse, the Moon is crossing the meridian. Corona, the Crown, is crossing the meridian and is in zenith position (the point in the sky directly above an observer) at approximately 12:00 midnight, Thursday, April 26.

Signs of His Resurrection: Firstfruits (Nisan 18/April 29, AD 31)

13. Resurrection morning, Sunday, April 29, AD 31, the Moon (God's faithful witness in the sky) passes the meridian at 3:00 a.m. in Sagittarius, the Archer, representing the Messiah's spiritual offspring.

14. Resurrection morning, Sunday, April 29, AD 31, Vega crosses the meridian with Lyra in zenith position at 3:12 a.m.

15. Resurrection morning, Sunday, April 29, AD 31, the image of the constellation Aries, the risen Lamb, reemerges in the eastern sky at 4:30 a.m.

16. Resurrection morning, Sunday, April 29, AD 31, the constellation Perseus, the Breaker (symbolizing Jesus), rises from "the dark

domain" at 4:30 a.m. with the Sun aglow at his feet, clutching (symbolically) Satan's head tightly in his left hand.

17. Resurrection morning, Sunday, April 29, AD 31, Ophiuchus, the Serpent Restrainer (representing the Messiah), is crushing Scorpio, the Scorpion (representing Satan), as it descends beneath the southwest horizon at 4:45 a.m.

18. Resurrection morning, Sunday, April 29, AD 31, Cygnus, the Swan (the Northern Cross), Delphinus, the Dolphin, and Capricorn, the Goat, appear at the meridian at 5:00 a.m. before disappearing into the sunlight.

Appendix 3

Constellation and Star Name Definitions

Summary in Hebrew

1. **VIRGO:** (Latin) "virgin" (Hebrew) *Bethulah*, "virgin." Messiah's virgin birth.

    1. Spica: (Latin) *spīca*, "ear of wheat"/(Hebrew) *tsemach*, "branch." Tsemach is a prophetic name of the Messiah, the glorious branch of *Yᵉhôvâh*, "LORD" (John 12:24; Isa 4:2).

    2. Zavijava: (Hebrew) *tsev·ē*, "beautiful," and *Yᵉhôvâh*, "LORD." The two words combined mean: "beautiful LORD or (gloriously) beautiful LORD" (Isa 4:2).

    3. Vindemiatrix: combines three (Latin) words, plus the suffix *trix*. The three words are (1) *vīnī*, "wine, figuratively, grapes, grapevine", (2) *dē* "from or out of," and (3) *emia* "suffix meaning blood." The three words combined = "wine (or grapes/grapevine) out of blood." Also known as "vine-harvester" or "grape-harvester."

    4. Subilon: (Hebrew) *shibbol* or *shibboleth*, "ear of grain, branch."

    5. Zaniah: (Hebrew) *azan*, "hears," and *Yᵉhôvâh*, "LORD." The two words combined = "the LORD hears."

    6. Syrma: (Latin) *syrma*, "a robe with a train."

2. **COMA:** (Hebrew) כָּמַהּ/*kâmah*, "to pine after:—long" and *chemdah*, "the desired, the longed for" (Hag 2:7); (Latin) *coma* "hair." "The Longed-For" (Messiah).

3. **CENTAURUS:** (Latin) "Centaur" Messiah's message is carried throughout the earth.

   1. Toliman: (Hebrew) *olam*, "forever, everlasting, eternal, ancient, always."

   2. Hadar: (Hebrew) *hâdâr*, "magnificence, splendor, beauty, majesty, glorious."

4. **BOÖTES:** (Greek) *Boötes* "herdsman, ox-driver." Yeshua the Good Shepherd and Ox Trainer.

   1. Arcturus: (Greek) *arktos*, "bear" and *oûros*, "guardian, watcher." The two words combined = "bear-watcher" or "guardian of the bear." Boötes is symbolically guarding and guiding "flocks and herds" not bears.

   2. Nekkar: (Hebrew) *niq·qar*, "pierces, are pierced" from *nâqar*, "pierced, dig." To "goad" produces a piercing effect. Arabic: Nekkar, "the prodder."

   3. Mizar: (Arabic) *mi'zar*, "apron," a corresponding Hebrew word for the word *mizar* (a type of apron) or *ephod*, meaning "a girdle, apron-like garment."

   4. Muphrid: (Arabic) "single," (Hebrew cognate) *parad*, "separated, to divide."

   5. Al Kalurops: (Arabian version of the Greek) *kalaurops*, "the shepherd's crook."

5. **LIBRA:** (Latin) "scales"/(Hebrew) Moznayim "scales." God's supreme justice, Yeshua, our righteousnes.s

   1. Zuben El Genubi: (1) (Akkadian) *zibānu*, "scale," (2) (Arabic) *el*, definite article "the," and (3) (Arabic) *janūb*, "south." The three words combined = "the scale south."

   2. Zuben Al Chemali: (1) (Akkadian) *zibānu*, "scale," (2) (Arabic) *al*, definite article "the," and (3) (Arabic) *samāl*, "north." The three words combined = "the scale north."

3. Zuben Akrabi: (1) (Akkadian) *zibānu*, "scale," and (Hebrew) (2) *qᵉrâb*, "conflict, war, scorpion." The two words combined = "scale of war."

4. Brachium: (Latin) *brachiōn*, "arm."

6. **CRUX:** (Latin) "cross" known as "the Southern Cross." <u>Messiah paid our sin debt.</u>

7. **LUPUS:** (Latin) "wolf" (Greek) *therion*, "wild animal," including cattle, a wild ram, or goat. The Hebrew equivalency of *therion* is *bᵉhêmâh*, which means "beast, cattle, sheep, goat, livestock (of domestic animals), wild beasts." <u>Messiah's sacrifice on our behalf.</u>

8. **CORONA:** (Latin) "crown." <u>Messiah has restored Adam's lost dominion.</u>

9. **SCORPIO:** (Latin) "scorpion." <u>Messiah's crushing blow to Satan's head, stung on the heel.</u>

   1. Antares: (Greek) *anti*, "against" and *ariēs*, "ram, a male sheep" = "against the Ram/Lamb."

   2. Lesath: (Persian) *las'at*, "scorpion sting," (Hebrew cognate) *lazuth*, "perverseness, perverse."

   3. Shaula: (Hebrew) *saw-lal*, "to exalt (self); reflexively, to oppose."

10. **SERPENS:** (Latin) "snake." <u>Satan's evil ambition thwarted by Messiah.</u>

    1. Unuk: (High Middle German) "snake." (Arabic) *unuq*, "neck."

    2. Cheleb: (Hebrew) *cheleb*, "fat, literally, or, figuratively."

    3. Alyah: (Hebrew) *âlâh*, "curse."

11. **OPHIUCHUS:** (Greek) *óphis*, "a snake" and *kátochos* "bearer, holder, restrainer, possessor." "serpent restrainer." <u>Yeshua restrained Satan from retaining Adam's crown.</u>

    1. Ras El Hagus: (Arabic) (1) *Ras*, "head," (2) *el*, definite article "the," and (3) Hagus from *ḥajaza*, meaning "restrain." The three words combined = "the head of him who restrains."

2. Mageros: (Greek) *mogerós*, "toiling, distressed."

3. Triophas: (Greek) *tropaios*, "of defeat," from trope (a rout) originally "a turning" (of the enemy).

4. Celbalrai: (Arabic) (1) *celb* from *qalb*, "heart," (2) *al*, definite article "the," and (3) *rai*, "shepherd." The three words combined = "heart of the Shepherd."

12. **HERCULES:** From the Greek words *hḗrōs*, "hero" and *kléos*, meaning "renown, glory." (Hebrew) Gibbôr, "strong, mighty, hero, champion." <u>Yeshua, the Mighty Warrior, crushed the dragon's head.</u>

1. Ras Al Gethi: (Arabic) (1) *Ras*, "head," (2) *al*, definite article "the," (3) Gethi from *Jathi*, "kneeling" (man) = "the head of the kneeler."

2. Kornephoros: (Greek) *korunē*, "club, mace, shepherds' staff" and *phoréas* or *foréas*, "bearer." The two words combined = "club bearer."

13. **SAGITTARIUS**: (Latin) "archer"/(Hebrew) Qesheth, "archer." <u>The Triumphant Archer shooting his arrows of truth at the heart of the scorpion.</u>

1. Naim (Namalsadirah): (Hebrew) (1) *naem*, "pleasant, delightful, sweet," (2) (Arabic) definite article *al*, "the," and (3) (Arabic) *sadirah* from *ssadr*, "chief, chairman." The three words combined = "the delightful chief."

2. Al Warida or (Hamalwarid): (1) (Arabic) definite article *al*, "the" and (2) (Arabic) *ḥamal*, "male lamb," (3) (Arabic) *warada*, "to arrive." The three words combined = "arrival of the lamb" or "the lamb arrives."

3. Nunki: (Sumerian) *Nun*, "prince" and *ki*, "earth" = "prince of the earth."

4. Al Nasl: (Arabic) definite article *al*, "the" and *nasl*, "offspring, descendants," "the offspring."

14. **LYRA**: (Latin) "lyre, harp." <u>Heaven's triumphant music.</u>

1. Vega: (Hebrew) possible derivative of *sâgâ*, "to grow, laud, magnify, extol with praise."

2. Sheliak or Shelyak: (Greek) from *chelóna*, "tortoise, turtle."
   The lyre had a convex back of tortoiseshell or made of wood,
   shaped like the shell.

3. Sulaphat: (Arabic) *sulaḥĭ*, "tortoise." Sulaphat relates to the
   type of lyre made from a tortoise shell.

15. **ARA**: (Latin) "a structure for sacrifice, altar." The altar of sacrifice.

16. **DRACO**: (Latin) "dragon" (Hebrew) (1) *Livyathan*, "serpent," "a sea
    monster or dragon" (2) *Aqallathon*, "crooked, twisted" (3) *Nachash*,
    "serpent." The three Hebrew words combined = "Leviathan the
    twisted serpent" (Isa 27:1). Satan's twisted deception destroyed by
    Messiah.

    1. Thuban: (Arabic) *ṭu'bān*, "snake."

    2. Ras Taban: (Arabic) *ras*, "head" *thuban*, "snake" or "head of
       the snake"; (also called) Al Waid: (Arabic) *al* is the definite
       article "the"/(Hebrew) *avad*, "to wander away, lose oneself; to
       perish, destroy."

    3. El Tanin: (Arabic) *el* is the definite article "the" and (Hebrew)
       tannin, "serpent, dragon, sea monster" (Exod 7:12).

    4. Giansar: (Hebrew) גַּן/*gan*, "garden" and שַׂר/*sar*, "prince" or
       "prince of the garden"

    5. El Atik: (Arabic) *el* is the definite article "the"/(Arabic/Persian)
       "the antique, old." (Hebrew) *âthaq*, "to remove, to grow old;
       specifically, to transcribe: copy out, leave off, become (wax)
       old, remove."

    6. Grumium: (Hebrew) *Ârûwm*, "cunning, crafty, subtle."

Appendix 4

Planet Definitions Including Earth, Sun, and Moon

Summary in Hebrew

1. EARTH: (Hebrew) אֶרֶץ/*eretz,* "earth, land."
   The earth is the Lord's and the fullness thereof (Ps 24:1, 115:16).

2. SUN: (Hebrew) שֶׁמֶשׁ/*shemesh,* "to be brilliant; the Sun."
   Yeshua, the greater light (the Sun of Righteousness), is the light of the world (Gen 1:16; John 1:4–9; John 8:12).

3. MOON: (Hebrew) לְבָנָה/*lᵉbânâh,* "properly, (the) white, i.e., the Moon, the white one."
   The Moon, the lesser light, represents Israel/the church reflecting the light of the Son (Yeshua) (Gen 1:16; Ps 89:37).

4. JUPITER: (Hebrew) צדק/*tzedek,* "righteousness, justice."
   Yeshua is the Lord our righteousness (right standing with God) (Jer 33:14–16; 2 Cor 5:21).

5. VENUS: (Hebrew) נֹגַהּ/*kokhevet, nogah,* or *kokhav-nogah,* "the bright one," from נָגַהּ/*nâgahh,* "to glitter; causatively, to illuminate:—(en-) lighten, (cause to) shine."
   Yeshua is the bright morning star and the true light that gives light to the world.

291

The planet Venus is rendered as "morning star" by the ancients. Yeshua is referenced as the Day-Star in 2 Pet 1:19 and is called the Bright Morning Star in Rev 2:28 and 22:16.

6. SATURN: (Hebrew) שבתאי/*shabtai*, "from shabbat (sabbath), the restful one."

    Yeshua gives his people rest (Isa 56:6–7; Matt 11:28–30; Heb 4:9).

7. MARS: (Hebrew) אָדַם/*âdam*, "to show blood (in the face), i.e., flush or turn rosy:—be (dyed, made) red (ruddy)." "The red one."

    In righteousness, Yeshua makes war on the enemy and redeems us to himself (Rev 19:11–21; 1 Pet 1:19).

8. MERCURY: (Hebrew) כּוֹכַב חַמָּה/*ko•chav cha•ma*, "the word *cha•ma* is the literary name of the Sun, the Sun Star."

    The Messiah is the covenant messenger. His word travels swiftly throughout the world (Mal 3:1; Ps 147:15; Mark 16:15).

Appendix 5

Dictionary, Lexicon Resources, and References

Hebrew and Greek

Blue Letter Bible: https://www.blueletterbible.org
Bible Hub: https://biblehub.com
Glosbe Hebrew Dictionary: https://glosbe.com
Equip God's People: https://www.equipgodspeople.com
Glosbe (Greek) Dictionary: https://glosbe.com/en/el/a
Strong's Greek and Hebrew Concordance and Lexicon: https://studybible.info/strongs/
Ancient Hebrew Research Center: https://www.ancient-hebrew.org/
Smith's Bible Dictionary: https://www.htmlbible.com/kjv3o/smith/
Bible Study Tools: https://www.biblestudytools.com/lexicons/
Perseus Digital Library: https://www.perseus.tufts.edu/hopper/definitionlookup?type=
    begin&q=&target=greek
Hebrew English Parallel Bible: https://www.wordproject.org/bibles/parallel/hebrew/
Greek English Parallel Bible: https://www.wordproject.org/bibles/parallel/greek/

Aramaic

COMPREHENSIVE ARAMAIC LEXICON PROJECT: https://cal.huc.edu
Aramaic Lexicon and Concordance (Aramaic): https://www.atour.com/dictionary
Study Light Lexicon (Aramaic): https://www.studylight.org/lexicons/aramaic/1.html
E-Sword Aramaic Bible: https://www.e-sword.net
Aramaic Glossary: http://doormann.tripod.com/aramaic.htm
The Original Aramaic New Testament in Plain English: https://buffaloriverforge.com/
    peshitta/NT%20Peshitta%207th%20Ed%20Plain%20Text%20Unnoted.pdf.

Arabic

Bab.La (Arabic): https://en.bab.la/dictionary/english-arabic/
TamilCube: http://dictionary.tamilcube.com/arabic-dictionary.aspx
Lexilogos (Arabic): https://www.lexilogos.com/english/arabic_dictionary.htm
Arabic Lexicon: https://www.lexicons.ru/modern/a/arabic/ar-en-a.html
Arabic English Parallel Bible: https://www.wordproject.org/bibles/parallel/arabic/

Semitic Roots

The American Heritage Dictionary Semitic Roots Appendix: https://ahdictionary.com/
    word/semitic.html

Egyptian Arabic

Egyptian Arabic Dictionary: https://www.egyptianarabicdictionary.com/info/en/
    home.html

Latin

Learn That (Latin) Roots: https://www.learnthat.org/pages/view/roots.html
Glosbe (Latin) Dictionary: https://glosbe.com/en/la/corona
Latin Lexicon.org: http://latinlexicon.org/search_english.php
Numen—The Latin Lexicon: https://latinlexicon.org/index.php
English-Latin Dictionary: https://www.online-latin-dictionary.com/
Latin English Parallel Bible: https://www.wordproject.org/bibles/parallel/latin/

Akkadian

Akkadian Dictionary: http://www.assyrianlanguages.org/akkadian/search.php

Assyrian

Assyrian Languages: http://assyrianlanguages.org/sureth/index.php
English to Assyrian Dictionary: http://sargonsays.com/

Persian

Old Persian: Base Form Dictionary: http://dsal.uchicago.edu/dictionaries/steingass/
Glosbe (Persian Dictionary): https://glosbe.com/en/fa
Steingass, Francis. A Comprehensive Persian-English Dictionary: https://dsal.uchicago.edu/dictionaries/steingass/
Hayyim, Sulayman. New Persian-English Dictionary: https://dsal.uchicago.edu/dictionaries/hayyim/
Lexilogos (Persian): https://www.lexilogos.com/english/persian_dictionary.htm
Persian English Parallel Bible: https://www.wordproject.org/bibles/parallel/farsi/

Sumerian

Sumerian Lexicon by John A. Halloran: http://www.sumerian.org/sumerlex.htm http://www.sumerian.org/sumerian.pdf
Sumer.Grazhdani.eu: https://sumer.grazhdani.eu/index.php

Sanskrit

Sanskrit Dictionary: http://sanskritdictionary.com/?q=
Learn Sanskrit Dictionary: https://www.learnsanskrit.cc/

Hindi

Bab La (Hindi): https://en.bab.la/dictionary/english-hindi/
Platts, John T. A dictionary of Urdu, classical Hindi, and English: https://dsal.uchicago.edu/dictionaries/platts/
Hindi ShabdKhoj: https://dict.hinkhoj.com/english-to-hindi/
Hindi English Parallel Bible: https://www.wordproject.org/bibles/parallel/hindi/

Turkish

Glosbe (Turkish Dictionary): https://glosbe.com/en/tr/a
Turkish English Parallel Bible: https://www.wordproject.org/bibles/parallel/turkish/

Phoenician

A Short Vocabulary of Phoenician: http://ancientroadpublications.com/Studies/AncientLanguage/Phoenician.pdf
Glosbe (Phoenician Dictionary): https://glosbe.com/en/phn

Tamil

Glosbe (Tamil Dictionary): https://glosbe.com/en/ta/a
University of Madras. Tamil lexicon: https://dsal.uchicago.edu/dictionaries/tamil-lex/
Tamil English Parallel Bible: https://www.wordproject.org/bibles/parallel/tamil/index.htm

Egyptian

Egyptian Hieroglyphic: https://archive.org/stream/egyptianhierogly01budguoft#page/434/mode/2up/search/a
English to Egyptian Dictionary: https://karathutmose.tripod.com/dictionary/dictionary1.html
Glosbe (Egyptian) Dictionary: https://glosbe.com/en/egy
Coptic Dictionary Online: https://coptic-dictionary.org/

Amharic

English—Amharic Dictionary: https://www.amharicpro.com/
Glosbe (Amharic Dictionary): https://glosbe.com/en/am
Amharic English Parallel Bible: https://www.wordproject.org/bibles/parallel/amharic/

Tigrinya

Tigrinya Dictionary: https://www.geezexperience.com/
Glosbe (Amharic Dictionary): https://glosbe.com/en/ti

Chinese

Chinese Dictionary: http://www.chinese-dictionary.org/
Chinese English Parallel Bible: https://www.wordproject.org/bibles/parallel/chinese/

Japanese

Japanese Dictionary: https://japanesedictionary.org
Glosbe (Japanese): https://glosbe.com/en/ja
Japanese English Parallel Bible: https://www.wordproject.org/bibles/parallel/japanese/

Language Specialist References

Dr. Nehemia Gordon: Nehemias Wall: https://www.nehemiaswall.com/

Jeff Benner: Ancient Hebrew Research Center: https://www.ancient-hebrew.org/
Abarim Publications: https://www.abarim-publications.com/
CGE Jordan Institute for Arabic Studies: https://cgejordan.com/

Calculators

Time and Date: https://www.timeanddate.com/date/duration.html
webmaster@timeanddate.com
BC & AD Date Calculator: www.calculateconvert.com/calculators/date/history.php

Astronomy Resources

The Star That Astonished the World by Earnest L. Martin: https://www.askelm.com/
   star/
The Star of Bethlehem by Frederick A. "Rick" Larson: https://bethlehemstar.com/
The Witness of the Stars by E. W. Bullinger
The Gospel in the Stars by Joseph A. Seiss
The Real Meaning of the Zodiac by D. James Kennedy
Mazzaroth by Frances Rolleston
Signs in the Heavens by Marilyn Hickey
Sting of the Scorpion by Jonathan Grey
Signs In The Heavens by Chuck Missler
Earth Sky: https://earthsky.org/
Space.com: https://www.space.com/
NASA: https://www.nasa.gov/
Astronomy for Kids: https://www.astronomy.com/tags/astronomy-for-kids/
David Rives Ministries: https://davidrivesministries.org/ https://wonderscenter.org/

Astronomy Software

Stellarium: https://stellarium.org
Starry Night / Simulation Curriculum: https://starrynight.com
NASA Eclipses: https://eclipse.gsfc.nasa.gov / https://solarsystem.nasa.gov/eclipses/
   home/

Bibliography

Allen, R. Hinckley. *Star Names and Their Meanings*. Glastonbury, UK: G. E. Stechert, 1899.

American Philosophical Society (vol. 209). Harmony of the World by Johannes Kepler. https://www.pennpress.org/9780871692092/harmony-of-the-world-by-johannes-kepler/

Bayer, Johann. *Uranometria*. Augsburg: Christoph Mang, 1603.

British Museum. *Lyre*. Wood and ceramic. 5th–4th century BC. London, England. https://www.bmimages.com/preview.asp?image=00256226001&itemw=0&itemf=0001&itemstep=1&itemx=1.

———. *Papyrus Weighing Scales*. Papyrus, ca. 1250 BC. London, England. https://www.britishmuseum.org/collection/object/Y_EA10470-3.

Brumley, Albert E. "I'll Fly Away." Hartford Music Company, 1932.

Bullinger, E. W. *The Witness of the Stars*. Grand Rapids: Kregel, 1984.

Chester, Craig. "The Star of Bethlehem." *Imprimis* 22:12 (Dec 1993). https://imprimis.hillsdale.edu/the-star-of-bethlehem/.

Definitions.net. "Beta Lyrae." https://www.definitions.net/definition/beta+lyrae.

Dix, William Chatterton. "Hallelujah! Sing to Jesus." Bristol, England, 1866.

Espenak, Fred. "NASA Eclipse Web Site." https://eclipse.gsfc.nasa.gov/ / https://science.nasa.gov/eclipses/ / https://www.mreclipse.com/

Fleming, C. Kenneth. *God's Voice in the Stars*. Neptune, NJ: Loizeaux Brothers, 1981.

"God Rest Ye Merry, Gentlemen." Traditional English carol. Roud Folk Song Index 394. Ca. 1760

Gordon, Nehemia. "Hebrew Voices #47—A Disastrous Misunderstanding of the Name Yehovah." *Nehemia's Wall*, Mar 11, 2020. https://www.nehemiaswall.com/disastrous-misunderstanding-yehovah.

Hickey, Marilyn. "Signs in the Heavens." Marilyn Hickey Ministries, Jan 1, 1984. Denver, Colorado. https://www.marilynandsarah.org/.

Josephus, Flavius. *The Works of Josephus*: Updated ed. Peabody, MA: Hendrickson, 1987.

Kennedy, D. James. *The Real Meaning of the Zodiac*. Fort Lauderdale, FL: TCRM, 1993.

Larson, Rick, dir. *The Star of Bethlehem*. DVD. Mpower Pictures and Stephen Vidano Films: United States, 2007. https://bethlehemstar.com.

Lehman, Frederick. "The Love of God." 1917. https://hymnary.org/text/the_love_of_god_is_greater_far.

Lewis, C. S. *Reflections on the Psalms*. New York: Harcourt, Brace, 1986.

Martin, L. Earnest. *The Star That Astonished the World.* 2nd ed. Portland, OR: ASK, 1996. https://www.askelm.com/star/.

Mote, Edward. "On Christ the Solid Rock I Stand." 1836. https://hymnary.org/text/my_hope_is_built_on_nothing_less.

NASA "Orbits and Kepler's Laws." NASA, Jun 26, 2008. https://science.nasa.gov/resource/orbits-and-keplers-laws/.

Peretti, Frank. "God's Way or My Way—Part 2: Interview with James Dobson." *Family Talk*, Dr. James Dobson Family Institute. https://dobsonlibrary.com/resource/article/00bcod1c-a5a6-442f-8bf4-590a68aaf33c.

Rolleston, Frances. *Mazzaroth, or, the Constellations.* London: Rivingtons, 1862.

Seiss, A. Joseph. *The Gospel in the Stars.* Grand Rapids: Kregel, 1972.

Smith, George, and Archibald Henry Sayce. *The Chaldean Account of Genesis.* Cambridge: Cambridge University Press, 1876. https://www.gutenberg.org/files/60559/60559-h/60559-h.htm.

Targum 1 Chronicles. Sefaria. https://www.sefaria.org/texts/Tanakh/Targum.

Wesley, Charles. "Amazing Love! How Can It Be." Epworth, Great Britain, 1738.

Index

www.ingramcontent.com/pod-product-compliance
Lightning Source LLC
Chambersburg PA
CBHW062339300326
41947CB00012B/443